Lecture Notes in Bioinformatics 6620

Edited by S. Istrail, P. Pevzner, and M. Waterman

Subseries of Lecture Notes in Computer Science

W0038568

Ana T. Freitas   Arcadi Navarro (Eds.)

Bioinformatics for Personalized Medicine

10th Spanish Symposium, JBI 2010
Torremolinos, Spain, October 27-29, 2010
Revised Selected Papers

 Springer

Series Editors

Sorin Istrail, Brown University, Providence, RI, USA
Pavel Pevzner, University of California, San Diego, CA, USA
Michael Waterman, University of Southern California, Los Angeles, CA, USA

Volume Editors

Ana T. Freitas
INESC-ID/Instituto Superior Técnico
R. Alves Redol 9
1000-029 Lisboa, Portugal
E-mail: atf@kdbio.inesc-id.pt

Arcadi Navarro
Institut de Biologia Evolutiva (UPF-CSIC)
Doctor Aiguader 88
08003 Barcelona, Spain
E-mail: arcadi.navarro@upf.edu

ISSN 0302-9743   e-ISSN 1611-3349
ISBN 978-3-642-28061-0   e-ISBN 978-3-642-28062-7
DOI 10.1007/978-3-642-28062-7
Springer Heidelberg Dordrecht London New York

Library of Congress Control Number: 2011945508

CR Subject Classification (1998): H.3, H.2.8, F.2.1, H.4, C.2, H.5, D.2

LNCS Sublibrary: SL 8 – Bioinformatics

Preface

This volume contains the papers presented at JBI 2010: the 10th Spanish Symposium on Bioinformatics held in Torremolinos, Málaga, Spain, during October 27–29, 2010. After 10 years of conferences (the JBI conference series was started in 2000) JBI 2010 centered on the topic "Bioinformatics for Personalized Medicine" and was hosted by the University of Málaga, co-organized by the National Institute of Bioinformatics, Spain and the Portuguese Bioinformatics Network chaired by Oswaldo Trelles, Arcadi Navarro, Federico Morán and Pedro Fernandes and took place in the Hotel Son Don Pablo, at Paseo Marítimo, Torremolinos.

This year, 18 papers were accepted for presentation out of more than 100 submissions. The papers presented were selected by the Program Committee (PC) assisted by a number of external reviewers. Each paper was reviewed by two members of the PC, or by external reviewers. JBI 2010 also introduced a Highlights Track, in which eight additional presentations were chosen from journal papers published in 2009 or 2010. This made a total of 26 presentations. Authors were invited to submit a revised version of their work to be considered for publication in *Lecture Notes in Bioinformatics* after a second reviewing process by independent reviewers. This volume contains the papers that were finally selected.

JBI 2010 invited several distinguished speakers as keynotes and for special sessions on Genomic Projects Design: Success Cases, Bioinformatics Challenges in Personalized Medicine, Bioinformatics Software Developments in Spain and Beyond, and Options and Priorities in Hands-On Training in Spain and Portugal. A special session to present the status of the ELIXIR ESFRI project also took place. Invited speakers and special sessions organizers included Ivo Gut (National Genome Analysis Center Barcelona), Sara Hunter (EMBL-EBI Hinxton), Julio Saez Rodriguez (EMBL Outstation - Hinxton, European Bioinformatics Institute), Victoriano Valpuesta Fernández (University of Málaga), Alfonso Valencia (Centro Nacional de Investigaciones Oncológicas), José Pereira Leal (Instituto Gulbenkian de Ciência), Fernando Martin-Sânchez (Instituto de Salud Carlos III), Pedro Fernandes (Instituto Gulbenkian de Ciência), Jose María Fernández (Centro Nacional de Investigaciones Oncológicas), Jose Ramon Valverde (Centro Nacional de Biotecnologia), Victoria Martín-Requena (Universidad de Málaga), Maximiliano García (Universidad de Málaga) and Allan Orozco (Centro Nacional de Investigaciones Oncológicas).

JBI 2010 was only possible through the dedication and hard work of many individuals and organizations. Special thanks go to the PC and external reviewers for helping to improve the quality of the conference program, and the Organizing Committee, namely its Chair Oswaldo Trelles, for hosting the conference and providing the administrative, logistic, and financial support. We also thank our sponsors, including the INB (Instituto Nacional de Bioinformática), Bitlab (Bioinformatics and Information Technology Laboratory), Universidad de Málaga, Instituto de Salud Carlos III, Portuguese Bioinformatics Network, Life Sequencing, Integromics, UNIA, SEBIOT and Combiomed. Without them the conference would not have been financially viable. Finally, we thank all the authors who contributed papers and posters, as well as the attendees of the conference for their enthusiastic participation.

May 2011   Anna Teresa Freitas
                                                              Arcadi Navarro

Conference Organization

Honorary Committee

José Antonio Griñán Martínez (President of Andalucía)
Francisco de la Torre (Mayor of Málaga, Spain)
Pedro Fernández Montes (Mayor of Torremolinos, Spain)
Adelaida de la Calle Martín (Chancellor, University of Malaga, Spain)

Scientific Committee

Chairs

Alfonso Valencia	Centro Nacional de Investigaciones Oncológicas
Ana Teresa Freitas	INESC-ID/Instituto Superior Técnico

Committee

M. Mar Albà	Universitat Pompeu Fabra
Patrick Aloy	Institute for Research in Biomedicine
Jose Antonio Álvarez	Universidad de Santiago de Compostela
Julio R. Banga	Instituto de investigaciones marinas de Vigo
Ugo Bastolla	Instituto de Investigaciones Biomédicas
Pedro Bernaola Galván	Universidad de Málaga
Miguel Blanca	Hospital Carlos Haya
Christian Blaschke	BioAlma
José Luis Caballero	Universidad de Córdoba
José María Carazo	Centro Nacional de Biotecnología
Pedro Carpena	Universidad de Málaga
João Carriço	Universidade de Lisboa
Robert Castelo	Universitat Pompeu Fabra
Pablo Chacón	Centro de Investigaciones Biológicas
Julio Collados Vides	Universidad Nacional Autónoma de México
Ana Conesa	Centro de Investigaciones Príncipe Felipe
Francisco Couto	Universidade de Lisboa
Javier De Las Rivas	Universidad de Salamanca
Ramon Diaz-Uriarte	Spanish National Cancer Centre
Joaquin Dopazo	Centro de Investigación Príncipe Felipe
Hernán J. Dopazo	Centro de Investigación Príncipe Felipe

Julián Dorado	Facultad de Humanidades Universidade da Coruña
Eduardo Eyras	ICREA and Universitat Pompeu Fabra
André Falcão	Universidade de Lisboa
Toni Gabaldon	Centre for Genomic Regulation
Gian Gaetano Tartaglia	Center for Genomic Regulation
Juan Antonio García Ranea	Universidad de Málaga
Francesco Gervasio	Spanish National Cancer Research Centre
Manuel J. Gómez	Centro de Astrobiología INTA-CSIC
M. Gonzalo Claros	Universidad de Málaga
Antonio Granell	Instituto de Biología Molecular y Celular de Plantas de Valencia
Hugo Gutierrez de Teran	Fundación Pública Galega de Medicina Xenómica
Michael Hackenberg	Universidad de Granada
Fyodor Kondrashov	Center for Genomic Regulation
Josep Lluis Gelpi	Barcelona Supercomputing Center
Paulino Gómez-Puertas	Centro de Biología Molecular "Severo Ochoa"
Nuria Lopez-Bigas	Universitat Pompeu Fabra
Juan Fernandez-Recio	Barcelona Supercomputing Center
Antonio Figueras	Instituto de investigaciones Marinas
Roderic Guigó	Center for Genomic Regulation
Roberto Marabini	Universidad Autónoma de Madrid
Antonio Marin	Universidad de Sevilla
Miguel Angel Medina Torres	University of Malaga
Jordi Mestres	Instituto Municipal de Investigación Médica
Francisco Montero	Universidad Complutense de Madrid
Antonio Morreale	Universidad Autónoma de Madrid
Andrés Moya	University of Valencia
Juan Muñoz	Universidad de Cordoba
Arcadi Navarro	Universitat Pompeu Fabra
Cedric Notredame	Centre de Regulacio Genomica
José L. Oliver	Universidad de Granada
José Luís Oliveira	Universidade de Aveiro
Arlindo Oliveira	INESC-ID/Instituto Superior Técnico
Modesto Orozco	Barcelona Supercomputing Center
Alejandro Pazos	Universidad de A Coruña
José Pereira-Leal	Instituto Gulbenkian de Ciência
Alberto Pascual Montano	National Center for Biotechnology
Florencio Pazos	National Center for Biotechnology
Antonio Pérez Pulido	Universidad Pablo de Olávide
Francisco Pinto	Universidade de Lisboa

David G. Pisano   Spanish National Cancer Research Centre
David Posada   Universidad de Vigo
Raul Rabadan   Columbia University
Pedro Reche   Universidad Complutense de Madrid
Cristian Robert Munteanu   Universidade da Coruña
Isabel Rocha   Universidade do Minho
Ana M. Rojas   Institute for Predictive and Personalized
    Medicine of Cancer

Armindo Salvador   Universidade de Coimbra
Kika Sánchez   University of Malaga
Javier Sancho   Universidad de Zaragoza
Ferran Sanz   Institut Municipal d'Investigació Mèdica
Cláudio M. Soares   Instituto de Tecnologia Química e Biológica
Javier Tamames   National Center for Biotechnology
Oswaldo Trelles   Universidad de Málaga
Ana Tereza Vasconcelos   Universidad Federal do Río de Janeiro
Enrique Viguera-Mínguez   Universidad de Málaga
ordi Villà-Freixa   Universitat Pompeu Fabra

Organizing Committee

Chairs

Pedro Fernandes   Instituto Gulbenkian de Ciência
Federico Morán   Universidad Complutense Madrid
Arcadi Navarro   Universitat Pompeu Fabra
Oswaldo Trelles   Universidad de Málaga

Committee

Maximiliano García   Universidad de Málaga
Johan Karlsson   Universidad de Málaga
Victoria Martín-Requena   Universidad de Málaga
Alfredo Martínez-López   Universidad de Málaga
José Manuel Mateos   Universidad de Málaga
Antonio Muñoz-Mérida   Universidad de Málaga
Javier Ríos Pérez   Universidad de Málaga
Manuel Gonzalo Claros   Universidad de Málaga
Juan Falgueras Cano   Universidad de Málaga
Enrique Viguera Mínguez   Universidad de Málaga
Allan Orozco   Centro Nacional de Investigaciones Oncológicas

JBI Past Editions

Edition	Date	Organizing Committee
I	June 14-16, 2000	Universidad Politécnica de Cartagena
II	30 May – 1 June, 2001	Universidad de Málaga
III	September 18-20, 2002	Centro de Investigación del Cancer, Universidad de Salamanca
IV	September 13-15, 2003	Facultad de Informática de la Universidad de Coruña
V	29 November- 2 December, 2004	Universitat Politecnica de Catalunya
VI	28 September-2 October, 2005	Centro Nacional de Investigaciones Oncológicas
VII	November 20-22, 2006	Universidad de Zaragoza
VIII	February 13-15, 2008	Centro de Investigaciones Príncipe Felipe
IX	November 2-6, 2009	Instituto Gulbenkian de Lisboa

Table of Contents

Statistical Significance for NGS Reads Similarities

Antonio Muñoz-Mérida, Javier Ríos, Hicham Benzekri, and Oswaldo Trelles

Computer Architecture Department, University of Malaga, Spain
{amunoz,jriosp,ortrelles}@uma.es, bhicham538@gmail.com

Abstract. In this work we present a significance curve to segregate random alignments from true matches in by identity sequence comparison, especially suitable for sequencing data produced by NGS-technologies. The experimental approach reproduces the random local ungapped similarities distribution by score and length from which it is possible to asses the statistical significance of any particular ungapped similarity. This work includes the study of the distribution behaviour as a function of the experimental technology used to produce the raw sequences, as well as the scoring system used in the comparison. Our approach reproduces the expected behaviour and completes the proposal of Rost and Sander for homology based sequence comparisons. Results can be exploited by computational applications to reduce the computational cost and memory usage.

Keywords: assembly reads, similarity, NGS.

1   Introduction

New massively parallel data acquisition technologies in life sciences, in particular Next Generation Sequencing (NGS) are literarily producing orders of magnitude more data to analyze. For example, Roche's 454-sequencing (www.454.com), an ultra-high throughput sequencing technology based on a pyrosequencing principle, can produce terabytes of data compared with previous sizes ranging in gigabytes. Similar results are produced by ABI SOLiD (www.appliedbiosystems.com) and Solexa Illumina (www.illumina.com) platforms. These mountains of data represents a serious challenge for bioinformatics applications, in the early stages of raw data processing (e.g. sequencing data assembly) (Seqman [1], CABOG [2], EULER [3], MIRA [4]). Since NGS is currently one of the most active areas of innovation a particular effort must be paid to monitor latest advances and the new upcoming sequencing technologies such us SMRT technology [5]; nanopores [6] and semiconductors [7].

In this type of applications a first "clustering" phase is intended for segmentation of the input data set into smaller groups of sequences (clusters) that have stringent similarity, which are potentially coming from the same longer original sequence. In a second assembly phase each such cluster is used as input for a multiple alignment of the sequences to create one or more contigs (consensus sequences). In all of these procedures it is very common the use of 'all versus all' comparisons strategies with

A.T. Freitas and A. Navarro (Eds.): JBI 2010, LNBI 6620, pp. 1–7, 2012.
© Springer-Verlag Berlin Heidelberg 2012

data loaded into memory to speed-up computation. Easily these approaches raise the limits of computing resources, both at CPU and memory capacities since this software is being used for much larger data than originally designed.

Once pairwise sequence comparison is performed the next question is "are the compared sequences similar enough to be considered as the same view of a particular genomic region under study? Decision is usually made based on the similarity level and the length of the matched sequences. Some previous studies to answer this question are available. For instance, Rost [8] derived a curve to segregate random matches from the significant ones in protein sequences.

However, the scenario analysed in Rost's study was focused on protein data sets, therefore is valid to raise the question about the consistence of this curve when used with NGS sequencing data. In this scenario, there is a profusion of reads sharing a high level of similarity since they come from the same individual or from a closely related organism. It is no difficult to observe that 30% of similarity in 80 residues – valid for proteins is not enough for NGS fragments.

In this work we present a method to reproduce the expected behaviour for the ungapped similarities distribution for NGS-read data. The rationale is based on the computation of all local similarities without gaps between unrelated sequences in order to obtain the probability distribution of random similarities between NGS-reads.

2   System and Methods

The statistical significance is a robust criterion to evaluate the importance of pairwise similarities between biological sequences [9][10][11][12]. The problem can be stated in the following simply statistical terms: given a local similarity between two sequences evaluate the probability for this similarity to belong to random hits.

The gapped similarities case was studied by Karlin and Altschul [9] who reproduce the exponential behaviour of similarity distribution expected theoretically, supplying a mathematical approach to evaluate line parameters (slope and intersection). Sander and Schneider [12] and Rost [8] reproduced the set of random alignment and established a threshold curve to separate random from true alignments in protein sequences.

In this work we reviewed the previous work aimed to reproduce a significance curve to distinguish random from true similarities. The main differences with Rost approach are:

1. The underlying information about the precedence of the sequences: in this case, sequences are indeed products of sequencing belonging to the same region of the same genome.
2. Sequences under study are composed of nucleotides instead of aminoacids, therefore identity methods perform better than those based on similarity.
3. Both procedures aim to identify high conserved regions between two sequences but in our case, only situations allowed in assembly are allowed (e.g. sequence that are completely included in another longer sequence, or two sequences that are overlapped by their ends.
4. When working with proteins, isolated inner regions in common could show conserved domains, but in our study these cases are rejected because reads cannot be split for the final assembly.

In addition, we work with the initial premise that the reads come from the same specie, even the same individual, thus the identity when two reads belong to the same region must be almost perfect, except repeat regions, allelic variations and sequencing errors.

As Rost did for his curve, we constructed datasets of no related sequences in order to reproduce the random behaviour for later assume the real significance. The experiment was performed using three different datasets:

- A set of artificial reads generated from genomic DNA coming from two different chromosomes in *Sacharomyces cerevisiae*. Two fragments of genomic from chromosomes 1 and 4 were extracted producing a set of 5117 reads in chromosome 1 and 34074 reads in chromosome 4. The global coverage was set up to 10.
- A set of reads from ESTs to avoid intergenic repeats. ESTs were selected excluding ESTs from the same region or related in any way. The chosen specie was *Solanum lycopersicum* and 32930 reads compose the reads set.

Instead of using real datasets, for a better and simpler quality control we have performed a random selection [13] as the main criteria for building the sequences datasets.

Datasets were created by generation of random start points and lengths between prefixed ranks to simulate reads coming from different NGS technologies such as Roche-454, Illumina, and Solid to complete all the spectrum of reads length.

Our procedure begins by comparing reads each other in order to find the most similar region between them. We refer the similarity to the complete overlap common in both reads, not only the matched region, that is, we extend the HSPs (high-scoring segment pairs) to the ends to reproduce the way in which two reads would be aligned to take part into a longer consensus sequence (contig) following a head-tail alignment typical in assembly. The final alignment score is calculated as the identity percentage along the complete overlap.

The HSP needs to be extended till the ends to calculate its score. In case of not having a good score in the HSP overlap, the sequences will be considered not to belong to the same contig, in spite of having a good HSP.

3   Results

To establish the significance curve, sequences belonging to the datasets were compared. Real sequencing datasets was not used because of the ignorance of the sequences location in the chromosomes so we could not determine what sequences were really not related in order to establish the random behaviour so we proceed only with the artificial datasets. In the case of genomic reads the distribution differs from the expected. There is a homogeneous spot distribution both under and above the significance curve; the spot accumulation over the curve represents the high amount of repeats in the intergenic regions even at length of 1000nt.

In the second experiment we excluded the intergenic regions repeats using only coding sequences. In the spots distribution (see Figure 2), the trend we can distinguish 3 clear parts: the first one stays fix at 100% for sequences shorter than 30 nts. The second part shows a decreasing logarithmic behaviour, whose slope parameter depends on the dataset-type (e.g. in coding sequences it is steeper than genomic sequences), and finally the similarity curve tends to a constant value round to 45% in coding sequences and 40% in genomic sequences.

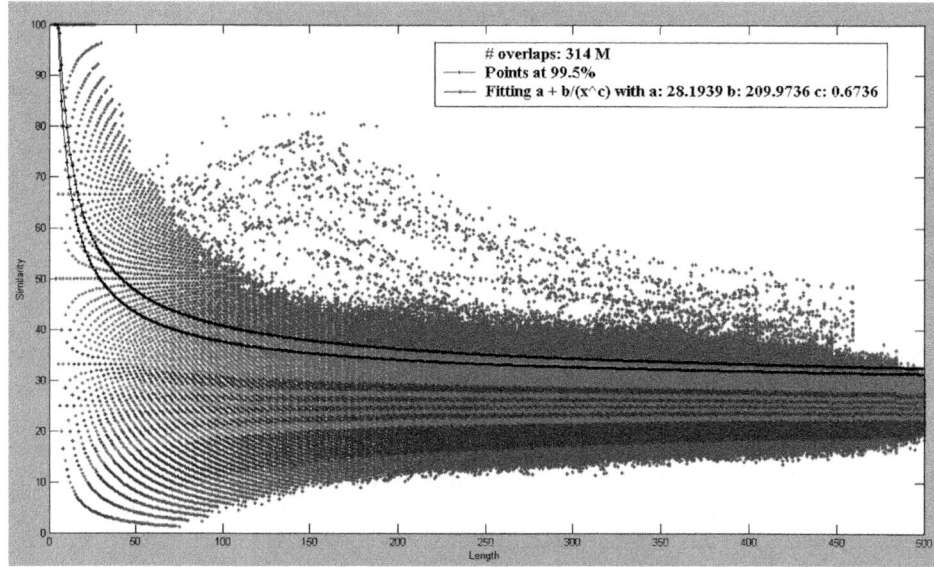

Fig. 1. Similarities distribution for artificial 454 Titanium genomic reads. Each band (dark/light grey) represents 10% of data. First and last band belong to 0.1% and 99.9%. Lines correspond to 99.5% and 99.9% of data for each length.

The significance curve was reproduced for different organisms such as mouse, olive and saccharomyces and technologies (454 FLX, 454 Titanium and Sanger) for setting a single equation valid for all of them looking at the homogeneity that all the graphics present.

Once the general equation was established, it was applied in a second process aimed to reduce the number of comparisons and therefore the CPU and memory requirements.

This process consists on a type of previous clustering to the final assembly to avoid the 'all vs all' comparisons that becomes out of control in huge datasets. The simple idea is compare one sequence with the rest of sequences and take all the sequences that have a higher score than the one marked for the significance curve for the overlap length in each case for their exclusion for later comparisons. Then the comparison will be continued with the first sequence in the list that has not been included in any group (see Figure 2). Reduction depends on the initial coverage and the genome repeats but with a couple of rounds we can reduce a lot the costs of CPU and memory that would allow the user to perform a bigger assembly, otherwise, it would be impossible without a computer upgrading.

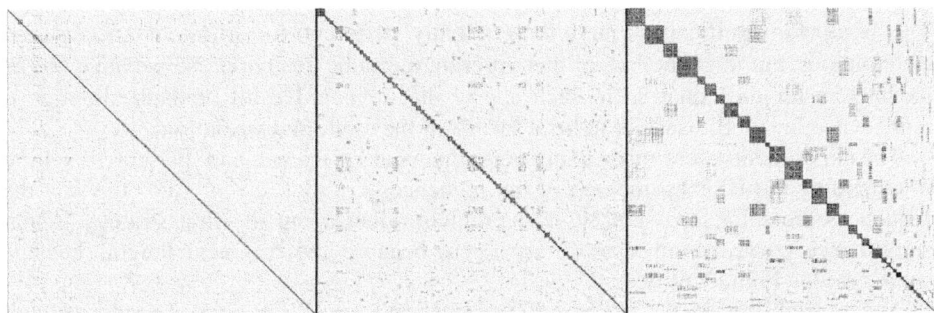

Fig. 2. Clustering of reads using the similarity threshold computed experimentally. Left square represents clusters in a 1x of coverage sequencing. In the middle, sequencing was done with a 10x of coverage, and the right one with a 50x of coverage. Numeric percentages of reduction are provided in Table 1.

In 'all vs all' method the total comparisons would be the half of the square area, but using the curve equation for grouping reads we only need to compare the horizontal lines from the small   squares till the end of the graph. The comparisons reduction can be seen at table 1:

Table 1. Computational space reduction for a N=1000 reads dataset with different coverage. The reference value (all-all comparison) is N*(N-1)/2.

	1x coverage	10x coverage	50x coverage
Calculated comparisons	290,875	69,034	14,319
% calculated	58.23	13.82	2.87
% reduction	41.77	86.18	97.13

4   Discussion

We propose a method to determine quickly whether two sequences obtained with any NGS technology must be assembled together or not.

Similar works were developed under the same idea, but taking into account a completely different scenario. Thus, most of estimations for pairwise sequence similarities have been done for proteins instead of nucleotides, and in case of nucleotides they include gaps. We present a study for a very different scenario that presents the following characteristics:

- Working with nucleotide sequences instead of protein sequences, and the target procedures are assembly processes.
- Sequence data comes from next generation sequencing (NGS) projects, which produces a broad range of sequences' lengths.
- It is expected to have sequences coming from the same region in a single organism or different organisms from the same species, thus the similarity level between sequences must be close to a full perfect match for most cases but SNPs, SSRs and sequencing errors.

The obtained significance curve was slightly different for different species and technologies, but in general all of them present the same behaviour, so we have set the parameters for the equation in such a way that it included all of them although it represented in some cases not to be at 99.9% of the random distribution.

The experiments were made using both, the local HSP score and the score resulting of extending the HSP till the end of the sequences in order to reproduce the way the sequences are joint in assembly. The problem when using the first strategy is that significance curve becomes more astringent because the fragment lengths became shorter with a high identity.

The availability of the proposed "segregation" curve for similar reads and our previous strategy [14] would allows to reduce the computational space (all-versus-all pairwise sequence comparison) and therefore the CPU and memory requirements in sequence assembly process. The experimental results with a set of 1000 reads shows that the 86% of the comparisons are not really needed to reproduce accepted results in a 10x coverage assembly. The process must be performed twice or three times to get the final clusters to be aligned as we can see in figure 2 at 50x of coverage because some relationships between groups can be reduced.

Acknowledgments. This work has been partially financed by the National Institute for Bioinformatics (INB) a platform of the ISCIII; the Spanish Integrated Action Programme (Code AT2009-0025); and by the European Science Foundation (ESF) - exchange grant activity Code 3064- in the programme Frontiers of Functional Genomics.

References

1. Swindell, S.R., Plasterer, T.N.: SEQMAN. Contig assembly. Methods Mol. Biol. 70, 75–89 (1997)
2. Miller, J.R., et al.: Aggressive assembly of pyrosequencing reads with mates. Bioinformatics 24, 2818–2824 (2008)
3. Pevzner, P.A., Tang, H., Waterman, M.S.: An Eulerian path approach to DNA fragment assembly. Proc. Natl. Acad. Sci. USA 98, 9748–9753 (2001)
4. Chevreux, B., Wetter, T., Suhai, S.: Genome sequence assembly using trace signals and additional sequence information. In: Comput. Sci. Biol.: Proc. German Conference on Bioinformatics GCB 1999 GCB, pp. 45–56 (1999)
5. http://www.pacificbiosciences.com/
6. http://www.nanowerk.com/news/newsid=17170.php#
7. http://www.technologyreview.com/biomedicine/23589/
8. Rost, B.: Twilight zone of protein sequence alignments. Protein Eng. 12(2), 85–94 (1999)
9. Karlin, S., Altschul, S.F.: Methods for assessing the statistical significance of molecular sequence features by using general scoring schemes. Proc. Natl. Acad. Sci. USA 87, 2264–2268 (1990)
10. Altschul, S.F., Gish, W.: Local alignment statistics. Methods Enzymol. 266, 460–480 (1996)
11. Collins, J.F., Coulson, A.: Significance of protein sequence similarities. Methods Enzymol. 183, 474–487 (1990)

12. Sander, C., Schneider, R.: Database of homology-derived protein structures and the structural meaning of sequence alignment. Proteins 9(1), 56–68 (1991)
13. Altschul, S.F., Bundschuh, R., Olsen, R., Hwa, T.: The estimation of statistical parameters for local alignment score distributions. Nucleic Acids Res. 29(2), 351–361 (2001)
14. Trelles, O., Andrade, M.A., Valencia, A., Zapata, E.L., Carazo, J.M.: Computational Space Reduction and Parallelization of a new Clustering Approach for Large Groups of Sequences. BioInformatics 14(5), 439–451 (1998)

Application of Array-Oriented Scientific Data Formats (NetCDF) to Genotype Data, GWASpi as an Example

Fernando Muñiz Fernandez[1,2], Angel Carreño Torres[1,2],
Carlos Morcillo-Suarez[1,2,3], and Arcadi Navarro[1,2,3,4]

[1] Institut de Biología Evolutiva (UPF-CSIC), Biomedical Research Park (PRBB), Barcelona
[2] Population Genomics Node (GNV8) National Institute for Bioinformatics (INB)
[3] National Genotyping Centre (CeGen)
[4] Institució Catalana de Recerca i Estudis Avançats, ICREA and Departament de
Ciències Experimentals i de la Salut,
Universitat Pompeu Fabra, Dr Aiguader 88, 08003 Barcelona, Spain

Abstract. Over the last three decades, the power, resolution and sophistication of scientific experiments has vastly increased, allowing the generation of vast volumes of biological data that need to be stored and processed. Array-oriented Scientific Data Formats are part of an effort by diverse scientific communities to solve the increasing problems of data storage and manipulations. Genome-wide Association Studies (GWAS) based on Single Nucleotide Polymorphism (SNP) arrays are one of the technologies that produce large volumes of data, particularly information on genomic variability. Due to the complexity of the methods and software packages available, each with its particular and intricate formats and work-flows, the analysis of GWAS confronts scientists with a complex hardware and software problematic. To help easing these issues, we have introduced the use of Array-oriented Scientific Data Format databases (NetCDF) in the GWASpi application, a user-friendly, multi-platform, desktop-able software for the management and analysis of GWAS data. The achieved leap of performance has permitted to leverage the most out of commonly available desktop hardware, on which GWASpi now enables "start- to-end" GWAS management, from raw data to end results and charts. Not only NetCDF allows storing the data efficiently, but it reduces the time needed to achieve the basic results of a GWAS in up to two orders of magnitude. Additionally, the same principles can be used to store and analyze variability data generated by means of ultrasequencing technologies. Available at http://www.gwaspi.org.

Keywords: Array-oriented Scientific Data Formats, NetCDF, HDF, GWAS, Genome wide association studies, single nucleotide polymorphisms, SNP.

1   Introduction

As standardized arrays from several vendors[1],[2] have become readily available, Genome Wide Association Studies (GWAS) have come to be the favored method to

A.T. Freitas and A. Navarro (Eds.): JBI 2010, LNBI 6620, pp. 8–20, 2012.

detect loci associated with human hereditary traits, specially diseases[3]. The number of studies published yearly based on these arrays, has constantly increased from 3 in 2003 to 384 in 2010[4]. In parallel, reference databases such as HapMap Phases II and III[5], HGDP[6] and the 1000 Genomes Project[7] are being made available together with an ever-growing range of analytical methods and software packages.

All of the extant tools, however, leave it to the user to tackle the jungle of formats and the bulk of raw data generated by GWAS. This burden of manipulation of the raw data still makes the access to GWAS a costly endeavor for departments without Bioinformatics personnel, often proving to be a persistent bottleneck right before publishing deadlines.

To contribute to solve this problem and make GWAS an achievable effort for smaller teams, as well as for the sake of speeding up raw data management in a consistent, self-contained way for the general researcher community, we have developed GWAS Pipeline (GWASpi). The main feature of GWASpi is that it implements NetCDF-3, an Array-oriented Scientific Data Format, which, along with a number of agile data-management methods, helps solving the data storage and access issues that are common to high throughput genotyping platforms.

GWASpi provides out-of-the-box quality control of raw data, typical frequency counts and Hardy-Weinberg threshold filters, basic allelic association tests over multiple phenotypes, chart generation and plot navigation as well as handy data management, such as import/export from/to a wide range of useful formats, data extraction, data merge and genotype translation from a range of encodings, all with easy to use, graphical wizards that guide the user through the process. The application creates a queue of tasks that is performed sequentially in a safe way. GWASpi can also be used through command-line interface, to use on servers without human interaction.

2   Features

2.1   Basic Analysis Features and Procedures

GWASpi is a cross-platform Java desktop-able application with a number of integrated tools and features that make the processing of genotype data more agile and speedy. All the features are accessible through a self-teaching, user-friendly, graphical interface while the heaviest functionalities can also be accessed via a command-line interface making use of script files.

The basic common usage starts by loading raw genotype data exported from a proprietary genotyping platform (see 2.2). Once the data has been imported to GWASpi's database, the next step is to perform the most common quality controls, such as missing genotypes per sample and per SNP as well as an allelic mismatch control.

Consequently, a genotype frequency count is performed, which will automatically include a Hardy-Weinberg quality control[8]. This step can be repeated multiple times with different phenotype data and/or quality control filter thresholds. Finally, once the frequency count is done, association tests can be performed, which will generate tables, reports and charts such as Manhattan and QQ-plots.

Result tables and charts can be viewed internally (Fig 1) or saved to an external file to be opened by standard editors. The internal results-table viewer also links each listed marker to reference databases such as Ensembl[9] and NCBI's dbSNP[10] among many others. A navigator is available to zoom inside a Manhattan-plot to view the detail of a chromosome area around a given marker, providing a first glimpse of the results of the GWAS as well as integration with the same external resources mentioned above.

All these steps can be performed on demand as well as in a convenient "GWAS in one go" command, with all necessary operations included sequentially. The application has a built-in thread queue management system which allows introducing a list of tasks to be performed when resources become available. The current desktop application's scope does not permit parallel processing due to the hardware's commonly available RAM configurations, but GWASpi's threading model has been written allowing for easy implementation of parallel usage. All queued processes and their logs are displayed so that monitoring and control is made easy.

As the user performs different operations, GWASpi keeps track of the work-flow and data it generates. All information and metadata of the loaded genotypes, sample information, quality controls and analysis, as well as the generated tables and charts, are stored in GWASpi's management database and are displayed inside a tree structure, commented with relevant descriptions.

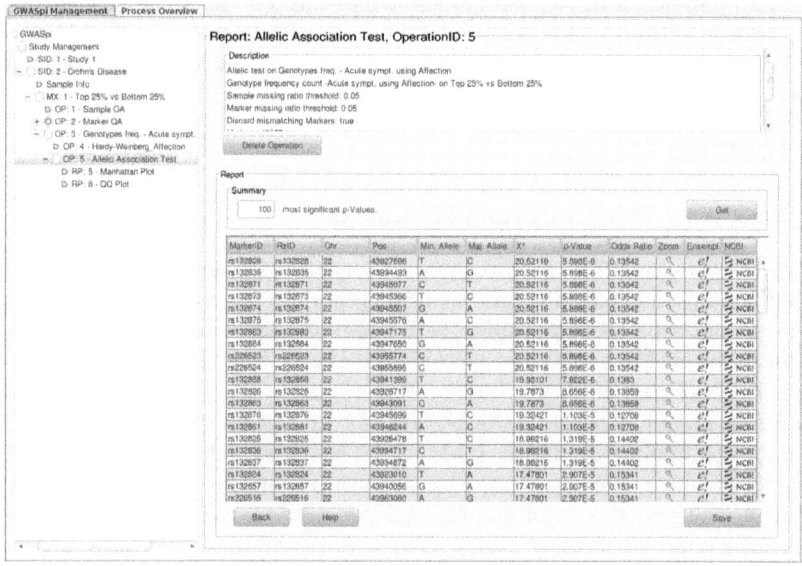

Fig. 1. GWASpi screenshot, Allelic Association report

Other GWASpi operations include a functionality to export data into the most common application formats (see 2.2). Additionally, an extraction manager allows the user to extract subsets of the original data to a new dataset by multiple criteria including or excluding markers by chromosome, marker ID and dbSNP ID (rsID). Simultaneously, samples can be included or excluded by sample ID, affection, age, category, disease, family ID, population and sex.

Merging of two datasets to a new one can be performed by either adding markers to a given set of samples (e.g. merging chromosomes to a single dataset) or adding samples to a given set of markers (e.g. merging several HapMap populations). Any extracted or transformed set of data is copied over to a new matrix so that the original dataset is preserved, ensuring non-destructive and safe processing.

2.2   Currently Admitted Formats

Import formats include, apart from GWASpi's own netCDF[11] files (see below), PLINK[12]), Affymetrix GenomeWide6[1], Illumina[2] (LGEN export format, based on PLINK), BEAGLE[13], HAPMAP[5] and HGDP[6] supplement1.

Export formats include, on top of GWASpi's matrix files, PLINK, transposed PLINK, BEAGLE and MACH[14], as well as a flat-text, header and index including matrix format, for easy work with spreadsheets and Perl scripts. The inclusion of new formats in GWASpi is straight-forward and more will be added as they are requested.

3   Performance

GWASpi, as a Java application, gets assigned a fixed size of the system's RAM, and will never request more than this quantity. In this way, the parameters of operation remain predictable and under control. The application has been designed to work on common desktop or laptop hardware, typically using 2GB of RAM per 10e6 SNPs with the number of samples being limited only by available disk-space.

3.1   RAM Allocation

By default, GWASpi is set-up to take 50% of the systems installed memory, but can easily be instructed to use more or less. The size of allocated memory determines how many SNPs the user will be able to load and process.

Table 1. Minimum RAM use benchmark for 3 real study cases

Number of SNPs	Allocated RAM [MB]
890.000	1000
2.610.000	3900
3.880.000	6000

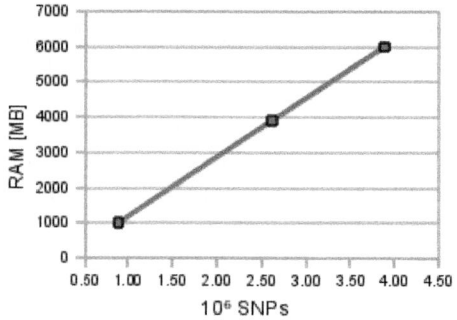

Fig. 2. Minimum RAM allocation per N° SNPs

3.2   Disk Usage

Sample size and SNP number will affect the number of the stored genotypes. GWASpi does not compress the genotype files as this feature is not available in netCDF-3. A genotype matrix with 2x10e6 SNPs and 1000 samples, as in the example above, occupies 4GB on disk (Fig 3). The disk-usage increase is proportional to the number of genotypes, which in turn is equal to SNPs number x sample size.

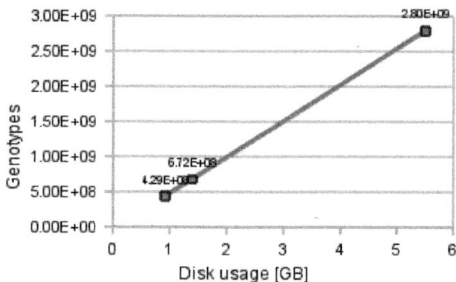

Fig. 3. Disk usage per genotype

3.3   Speed

The longest process the user will face while using GWASpi is the loading of a dataset into the GWASpi database. Depending on the format and the number of genotypes to be imported the loading of a dataset into the GWASpi can take as little as a few minutes (using GWASpi's own format) up to hours. As a guideline, a study with 900.000 SNPs and 1000 samples in Affymetrix GW 6.0 format takes a total of about 4 hours to load the data and perform quality control, Hardy-Weinberg and association tests as well as producing the corresponding reports and charts.

4   Implementation

The principal novelty included in GWASpi, the array-oriented scientific data storage paradigm, is mainly available in two technologies, HDF and NetCDF, of which the latter has been chosen for its pure Java implementation. The conjunction of this database technology together with a suite of high-performance libraries and tools (JfreeChart[15], Apache Derby[16], NetCDF-Java[11]) constitute the basis for GWASpi's development. They all are cross-plaform, easy to maintain and distribute, well established and tested. The mentioned key feature (more precisely the Java NetCDF-3 API) makes high-throughput processing accessible on standard desktop hardware at the scales necessary for GWAS.

4.1   Genotype Data Characteristics

The properties of genotypes can be described, in general terms, as follows:

- The data is dense (no sparse datasets, little missing data)
- Data doesn't change after acquisition
- The size of information of each array element is fixed and known (alleles)
- Genotypes can be ontologized and fit into custom data-objects (arrays and matrices)
- The order of genotypes follows convened rules (e.g. chromosomes, physical position)
- Genotypes can be indexed and referenced by unique IDs (dbSNP ID and/or proprietary IDs, sample ID)

Considering these properties, a convenient array-oriented scientific data storage system can be designed to fit our needs.

4.2   Storage Format Specifications

Typical GWAS genotype datasets present properties that make their storage and processing computationally expensive. Their scale typically ranges between 10e5 up to 4x10e6 SNPs, and between hundreds to tenths of thousands of samples. This storage and particularly the posterior retrieval of specific subsets of this data, at arbitrary places and times, is a highly desirable feature for the types of analysis that are being performed on these datasets. Generally, this type of functionality is achieved through the use of Relational Databases (RDB). Nevertheless, RDB technologies have poor scaling properties for the type of data that composes genome wide studies. Sequential binary formats used by platforms such as PLINK allow fast processing, but they are custom-made for the given application. The algorithms these formats permit are bound to the shape in which data is delivered on read access, be it ordered by samples or by markers.

GWASpi uses an Apache Derby RDB for storing sample information and application specific data such as work-flows but genotype subsets cannot be stored and retrieved in an agile way with the same technology. This is why sequential access of genotype data in flat-text files, not random-access, has been the prevailing methodology for many of the current GWAS applications.

Such sequential access methodologies, as parsing flat-text files, limit the type of operations that can be carried-out within a given time and on a given machine. They also precondition the algorithms that can be used to process the data, resulting in sub-optimal applications.

There are other scientific areas that have been confronted to similar problems since the 1980's and 1990's and to-date, a number of technologies have been developed and widely adopted. Parallelization and distributed storage and processing[17] are two examples, but they require a whole new layer of hardware and management complexity which would be difficult to implement for the scope of the application that we wanted to develop[18].

Other strategies have been identified and assessed to suit our storage and retrieval needs in GWAS analysis.

4.3   Array-Oriented Scientific Data Formats

Origins: The meteorological community at UCAR created the NetCDF format for storing satellite images as well as weather and geographical data in coordinate systems. In parallel, the particle collisioners around the world were also generating huge amounts of data that had to be stored and then processed. Researchers from NCSA, the Lawrence Livermore National Laboratory (LLNL), Sandia National Laboratory (SNL) and Los Alamos National Laboratory (LANL) developed a format called Hierarchical Data Format (HDF) to assist in these tasks. Both technologies were related in concept and ended up converging into what basically can be described as formats aimed at storing large multidimensional arrays of scientific data, with a focus on scalability. They both offer libraries in the most common programming languages, providing I/O and management methods that can be included in stand alone or client-server applications. These technologies store a description of data within the files themselves. They standardize the format and descriptions of many types of commonly used scientific data sets. They are also platform independent and the files are self-describing. For each data object they contain, there are predefined tags that identify such information as the type of data, the amount of data, its dimensions, and its location in the file. The self-describing capability of these files has important implications for processing scientific data. It makes it possible to fully understand the structure and contents of a file just from the information stored in the file itself.

One downside with these technologies is that one looses the ability to query the datasets with SQL. To circumvent this loss, the design of these databases has to include indexes, metadata and hierarchical structures to know where all its parts are located. In GWASpi's case, an index of samples and SNP markers is being constructed as the user loads genotype data in the system.

4.4   GWASpi's NetCDF-3 Database

The genotype data is stored using the specific methodology detailed below. GWASpi generates metadata that describes the data it contains. It also stores global attributes that apply to all its content:

Dimensions

–   Sample-Set-Dimension. Set to UNLIMITED. This means that this dimension can be set to an arbitrarily large number that depends only on the physical size of the hard-drive. The size of each element (ID) in this set is limited to 64 characters.

–   Marker-Set-Dimension. Sized to the number of markers provided in the dataset. There is an upper limit to the number of markers for each sample, namely the maximum unsigned integer number, or 4.294.967.295 SNPs. Beyond this number, a new matrix would have to be generated to store further data. The size of each element (ID) in this set is limited to 64 characters.

–   Genotypes-Matrix-Dimension. It will be the result of the "Marker-Set-Dimension" × "Sample-Set-Dimension". In other words, this will give shape to a matrix with "Sample-Set-Dimension" rows with "Marker-Set-Dimension" columns. The size of each cell ("Genotype-Dimension") in these matrices is currently limited to 2 bytes, one for each allele.

Table 2. Example of multidimensional matrix for genotype storage in GWASpi. Genotypes encoded as ACGT, in binary ASCII 910

		Marker-Set-Dimension						
		rs10001	rs10002	rs10003	rs10004	rs10005	rs10006	rs10007
Sample-Set-Dimension	smpl01	{65,84}	{84,84}	{71,71}	{65,67}	{71,71}	{65,84}	{71,71}
	smpl02	{65,65}	{84,84}	{71,71}	{65,67}	{71,84}	{65,65}	{71,71}
	smpl03	{65,65}	{84,84}	{71,71}	{65,67}	{71,71}	{65,65}	{71,71}
	smpl04	{65, 84}	{84,84}	{71,71}	{65,67}	{48,48}	{65, 84}	{71,71}
	smpl05	{65,65}	{67,84}	{71,71}	{65,67}	{71,71}	{65,65}	{71,71}
	smpl06	{65,65}	{84,84}	{71,71}	{65,67}	{71,84}	{65,65}	{71,71}
	smpl07	{65,84}	{67,84}	{71,71}	{65,67}	{71,84}	{65,84}	{71,71}
	smpl08	{84,84}	{84,84}	{71,71}	{65,67}	{71,71}	{84,84}	{71,71}
	smpl09	{65,65}	{84,84}	{71,71}	{65,67}	{71,71}	{65,65}	{71,71}
	smpl10	{65,84}	{84,84}	{71,71}	{65,67}	{71,71}	{65,84}	{71,71}

Attributes

–   Genotype-Encoding. This attribute specifies the encoding the genotypes are stored in (ACGT, 1234, AB, 12, UNKNOWN).

–   Technology. An attribute field to specify the original technology or source format the genotypes were provided in (PLINK, Hapmap, Affymetrix, Illumina...).

–   Description. User generated description of the matrix.

Metadata Arrays

–   Marker-ID. This ordered array of dimension "Marker-Set-Dimension" contains all the marker-IDs in a specific order (chromosome first, then physical position). The marker-IDs may be equal to the dbSNP database ID (rsID), but can be different if the original technology specifies it (e.g. Affymetrix).

- Rs-ID. An ordered array of dimension "Marker-Set-Dimension". This is dbSNP's rsID code for every marker, when available. It may be equal to marker-ID array.
- Chromosome. Array used for indexing and sorting, of dimension "Marker-Set-Dimension". This is the stated chromosome location of each marker. Chromosomes will be encoded and ordered as follows: 1-22, X, Y, XY and MT. XY means pseudo-autosomal, MT means mitochondrial.
- Position. Array used for indexing and sorting of dimension "Marker-Set-Dimension". This is the stated physical position within a chromosome of each marker (as per the annotation in the raw data). This position typically refers to a specific build of dbSNP's reference database.
- Encoding-Dictionary. An ordered array of dimension "Marker-Set-Dimension". This will hold a dictionary for translating a given encoding (AB or 12) to the dictionary's counterpart. Only provided and necessary for specific technologies (e.g. Affymetrix).
- Sample-ID. An ordered array of dimension "Sample-Set-Dimension". This is an ordered array of sample-IDs, as provided by the user. The order will be the one provided or the timing order in which the individual genotype files have been read in (depending on the input format).

Genotype-Matrix. The genotype matrix, as stated above will be a 3 dimensional matrix (Table 2), of dimensions "Marker-Set-Dimension" × "Sample-Set-Dimension" × "Genotype Dimension". The latter dimension is just a size specification of a genotype, in our case 2, one for each allele.

4.5   Writing Data to a Matrix

Every time a new matrix is created, data is written as specified above. NetCDF-3 is particular in the way it writes the data as one has to define first the shape (dimensions) of the data one will be writing, in what is called "Design Time". Once that phase is done, the shape is set and cannot be modified. GWASpi takes care of this step by initially ascertaining the shape and dimension of the provided data.

Next, during the "Write Time" phase, genotype data is written to the matrix. Once that phase is completed, the NetCDF-3 file is wrapped up and closed. After this the file cannot be edited or updated. This may seem like a major downside of NetCDF-3 but in our case, as genotype data remains static, it does not affect subsequent processes. Any later editing or processing, any operation or analysis that has to be performed on this matrix will yield data that will be written to a new NetCDF-3 file, always keeping the original untouched. This comes quite handy for keeping track of work-flows and having raw data backed-up. In a nutshell, any operation performed in GWASpi, short of a deletion, is non-destructive.

4.6   Retrieving Data from a Matrix

The principal task performed by the NetCDF-3 databases in GWASpi is on demand access to any fragment of data it contains. One may read rows or columns of the matrix as predefined subsets, regardless of their position inside the dataset, labeled with the marker and sample headers. This is done by using NetCDF-3's API and Java's LinkedHashMap (LHM) objects, which permit to program a fast and efficient way of manipulating arrays of data.

LinkedHashMaps (LHM) are ordered HashMaps, which combine the power of fast recovery of elements within an array, using a key, with a predictable sort order. The key will return its associated value within the LHM. Also, contrary to traditional HashMaps, the order of key/value pairs is specified and known. In this case, it will follow the order indicated in the reference marker-ID set and sample-ID set, as stored in the matrix metadata.

4.7   Other NetCDF-3 Databases Used in GWASpi

The above concept is also applied to other datasets that are generated by GWASpi, such as quality assurance, analysis data, reports and charts. In the case of quality assurance and analysis data, the test results are stored in character and numerical 1, 2 and 3 dimensional arrays, as needed, along with their correct shapes, dimensions and metadata, so as to be able to retrieve the context the data has been generated for. As before, LHMs are used to read/write to these databases.

5   Applications, Use Cases and Comparison

GWASpi has been used by the Spanish National Genotyping Center (CeGen www.cegen.org) to manage the data of several projects, ranging from studies on Multiple Sclerosis to the study of the evolution of complex traits during studies of association and genetic population stratification. Due to the recent release date of GWASpi, these studies are still pending publication.

Below we showcase a typical GWAS with 1200 individual samples, each genotyped with the Affymetrix GenomeWide 6.0 technology which holds 934968 polymorphic markers per chip. We depict the work-flow for typical preprocessing steps using Perl scripts, PLINK for the analysis and JfreeChart based plotting scripts for graphics (Fig 4a) as well as the same process performed with GWASpi (Fig 4b).

The steps depicted in dark grey are the bottlenecks of this work-flow: filtering unwanted and duplicate markers, sorting raw data files to ensure a pre-defined mapping of your genotypes and merging raw genotypes in a specific format. They may be performed in parallel on a cluster and resulting execution times will be reduced significantly, but nevertheless, the total CPU time required by these steps are a major hurdle in the GWAS process.

Intermediate consistency controls are crucial during these steps, as raw genotype data files are particularly difficult to predict as to its content. Assuming erroneous grammars of input formats is a common caveat and methods must be put in place to guarantee that data is processed and mapped correctly. This additional pre-processing time has not been accounted for in the depicted work-flow.

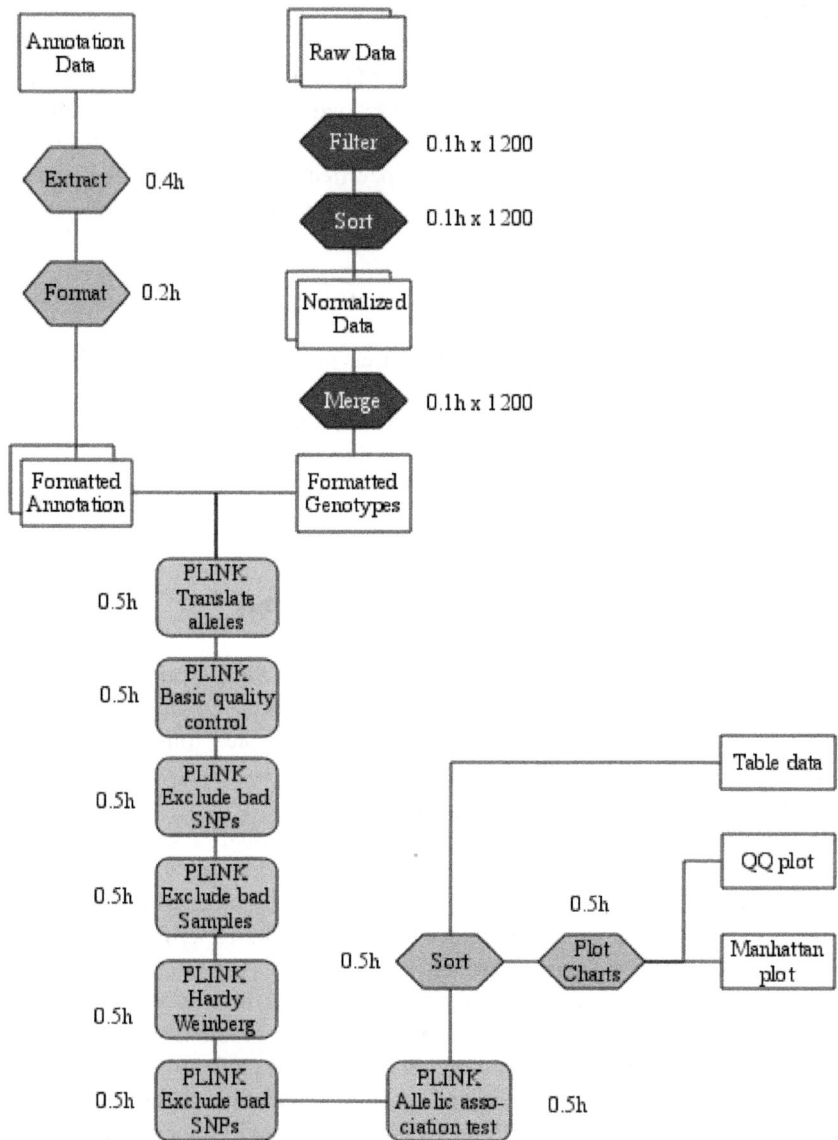

Fig. 4a. Workflow for 1200 individual samples genotyped on Affymetrix GW6.0, with 9x10e5 markers. Typical execution times on standard desktop hardware[1] shown, totaling 364.5 CPU hours (approx. 2 weeks). Not included are intermediate consistency checks.

---

[1] Intel Core2 Duo E4600, 2400GHz with 2.4GB RAM

The use by GWASpi of LHM's and scientific data arrays allows for a de-facto consistent treatment of genotypes, as the sorting, de-duping and mapping is done on the fly. Merging and writing of datasets is reduced to a single operation under GWASpi's database system. Not only does the application enforce strict consistency independently of the sort order and sparsity of the input files but it does so in a significantly shorter processing time, reducing it by two orders of magnitude in cases as the one mentioned above.

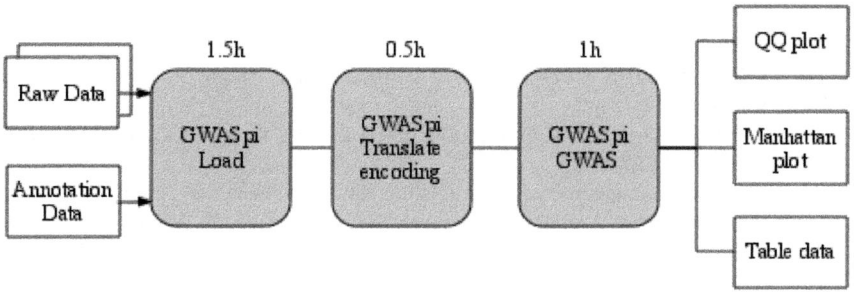

Fig. 4b. As above, using GWASpi. Typical execution times on standard desktop hardware shown, totaling 3 CPU hours. Consistency checks are built-in.

6   Conclusions

A case by case assessment of large experimental datasets common in bioinformatics, such as genotype studies and data coming from new-generation sequencing (NGS) technologies, considering them under the light of array-oriented scientific data formats originating from diverse fields of science, may open-up a whole new range of methods and algorithms that offer a wealth of optimization opportunities and a flexibility that was before sacrificed for lack of speedy data retrieval. For instance, in the case of NGS applications, the usage of the HDF-5 format would allow for compressed formats along with quick access and multidimensional storage spaces for agile shaping of data objects necessary for high throughput, parallel processing and networking applications. The use of netCDF has proven to be a significant improvement on the current data formats and management strategies in the case of GWASpi, a powerful hub platform upon which it has been easy to build light-weight, flexible and scalable applications and statistical methods that are ready for future increases of GWAS dataset size. New features, processing algorithms and import and export formats are easy to implement within its modular API structure.

For the user, GWASpi simplifies drastically the cumbersome work-flows so common in GWAS pipelines and the many intermediate data-files to be generated, checked and re-processed sequentially. The learning curve is kept shallow and its usage is very easy as all available operations are documented in an online contextual help web putting at your reach the know-how accumulated in a wealth of GWAS performed to-date.

References

[1] Yap, G.: Affymetrix, Inc. Pharmacogenomics 3(5), 709–711 (2002)

[2] Steemers, F.J., Gunderson, K.L.: Illumina, Inc. Pharmacogenomics 6(7), 777–782 (2005)

[3] McCarthy, M.I., Abecasis, G.R., Cardon, L.R., Goldstein, D.B., Little, J., Loannidis, J.P.A., Hirschhorn, J.N.: Genome-wide association studies for complex traits: consensus, uncertainty and challenges. Nature Reviews Genetics 9(5), 356–369 (2008)

[4] Yu, W., Wulf, A., Yesupriya, A., Clyne, M., Khoury, M.J., Gwinn, M.: HuGE Watch: tracking trends and patterns of published studies of genetic association and human genome epidemiology in near-real time. European Journal of Human Genetics EJHG 16(9), 1155–1158 (2008)

[5] International Hapmap Consortium: The International HapMap Project. Nature 426, 789–796 (2003)

[6] Luca Cavalli-Sforza, L.: The Human Genome Diversity Project: past, present and future. Genetics 6, 3–10 (2005)

[7] Siva, N.: 1000 Genomes project. Nature Biotechnology 26(3), 256 (2008)

[8] Gomes, I., Collins, A., Lonjou, C., Thomas, N.S., Wilkinson, J., Watson, M., Morton, N.: Hardy-Weinberg quality control. Annals of Human Genetics 63(pt. 6), 535–538 (1999)

[9] Ensembl, http://www.ensembl.org/

[10] dbSNP: the NCBI database, http://www.ncbi.nlm.nih.gov/SNP

[11] NetCDF at UCAR, http://www.unidata.ucar.edu/software/netcdf/

[12] Purcell, S., Neale, B., Todd-Brown, K., Thomas, L., Ferreira, M.A.R., Bender, D., Maller, J., Sklar, P., De Bakker, P.I.W., Daly, M.J., Sham, P.C.: PLINK: a tool set for whole-genome association and population-based linkage analyses. The American Journal of Human Genetics 81(3), 559–575 (2007)

[13] Browning, S.R., Browning, B.L.: Rapid and accurate haplotype phasing and missing-data inference for whole-genome association studies by use of localized haplotype clustering. The American Journal of Human Genetics 81(5), 1084–1097 (2007)

[14] Nothnagel, M., Ellinghaus, D., Schreiber, S., Krawczak, M., Franke, A.: A comprehensive evaluation of SNP genotype imputation. Human Genetics 125(2), 163–171 (2009)

[15] JFreeChart, http://www.jfree.org/jfreechart/

[16] Apache Derby, http://db.apache.org/derby/

[17] Sugawara, H.: Trends in bioinformatics. Tanpakushitsu Kakusan Koso Protein Nucleic Acid Enzyme 49(1), 72–73 (2004)

[18] Schadt, E.E., Linderman, M.D., Sorenson, J., Lee, L., Nolan, G.P.: Computational solutions to large-scale data management and analysis. Nature Reviews Genetics 11(9), 647–657 (2010)

[19] The HDF Group, http://www.hdfgroup.org/

An Ontology and a REST API
for Sequence Based Microbial Typing Data

João Almeida[1,*], João Tiple[1,*], Mário Ramirez[2], José Melo-Cristino[2],
Cátia Vaz[1,3], Alexandre P. Francisco[3,4], and João A. Carriço[2]

[1] DEETC, ISEL, Poly Inst. of Lisbon
[2] IM / IMM, FM, Univ. of Lisbon
[3] INESC-ID Lisbon
[4] CSE Dept., IST, Tech. Univ. of Lisbon

Abstract. In the Microbial typing field, the need to have a common understanding of the concepts described and the ability to share results within the community is an increasingly important requisite for the continued development of portable and accurate sequence-based typing methods. These methods are used for bacterial strain identification and are fundamental tools in Clinical Microbiology and Bacterial Population Genetics studies. In this article we propose an ontology designed for the microbial typing field, focusing on the widely used Multi Locus Sequence Typing methodology, and a RESTful API for accessing information systems based on the proposed ontology. This constitutes an important first step to accurately describe, analyze, curate, and manage information for microbial typing methodologies based on sequence based typing methodologies, and allows for the future integration with data analysis Web services.

Keywords: ontology, knowledge representation, data as a service, microbial typing methods.

1 Introduction

Microbial typing methods are fundamental tools for the epidemiological studies of bacterial populations [7]. These techniques allow the characterization of bacteria at the strain level providing researchers important information for the surveillance of infectious diseases, outbreak investigation and control, pathogenesis and natural history of an infection and bacterial population genetics. These areas of research have a direct impact in several human health issues, such as in the development of drug therapies and vaccines [1], with the concomitant social and economical repercussions.

With the decreasing cost and increasing availability of DNA sequencing technologies, sequence based typing methods are being preferred over traditional molecular methodologies. The large appeal of sequence-based typing methods is the ability to confidently share their results due to their reproducibility and

---

[*] These authors contributed equally to this work.

A.T. Freitas and A. Navarro (Eds.): JBI 2010, LNBI 6620, pp. 21–28, 2012.

portability, allowing for a global view and immediate comparison of microbial strains, from clinical and research settings all over the world. Several online microbial typing databases have been made available for different methods. The most successful examples are the Multi-Locus Sequence Typing (MLST) [6] databases for a multitude of bacterial species [10,12,8], *emm* typing database for *Streptococcus pyogenes* [14] and *spa* typing for *Staphylococcus aureus* [13].

However, these efforts are not standardized for data sharing, suffering from several caveats, being the most notable the lack of interfaces for automatic querying and running analysis tools. The automatic integration of data from the different databases is also hindered due to the lack of common identifiers among different databases. Moreover, the absence of an automatic validation of the new data in the submission process is leading to an increase of incomplete and unreliable data in the majority of these databases, seriously hampering the promised advantages of methodological accuracy and portability of results between laboratories. This is even more significant with the rise of new Single Nucleotide Polymorphism (SNP) typing techniques based upon the Next Generation Sequencing (NGS) [4] methods. The validity of this new high-throughput technology can be seriously hampered if the complete data analysis pipeline cannot be fully described in public databases, in order for the results to be reproducible. Also, the ability to integrate information from several well established typing methodologies will be paramount for the validation and development of the more informative whole genome approaches [5,3] based on these NGS methods for the bacterial typing field.

In a largely descriptive science such as Microbiology, the need to have a common understanding of the concepts described is fundamental for continued development of sequence-based typing methods. Therefore, the definition of an ontology that can validate and aggregate the knowledge of the existing microbial typing methods, is a necessary prerequisite for data integration in this field. In order to solve those problems, we present in this paper the design and implementation of an ontology created for the microbial typing field and an Application Programming Interface (API) to an information system using the concepts of the REST (Representational State Transfer) paradigm [2]. The proof-of-concept prototype of the proposed framework, focusing on the well established MLST methodology, is available at `http://rest.phyloviz.net`.

The ability to accurately describe the relationships between typing methods through the use of an ontology and to offer REST services to analyze, curate, and manage the information will facilitate the implementation of information systems capable of coping with the heterogeneous types of data existing in the field, including the re-usage of legacy data formats and methods.

This paper is organized as follows. Section 2 describes the proposed ontology, TypOn. Section 3 presents a REST API suitable for managing microbial typing data. Section 4 briefly details the RESTful Web services prototype implementation. Finally, Section 5 provides some final remarks and future work directions.

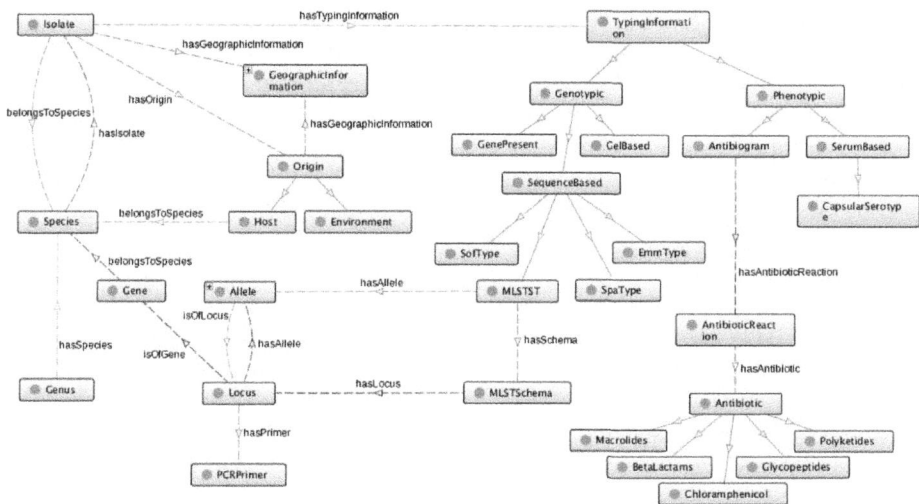

Fig. 1. TypOn, an ontology for microbial typing data. Dashed lines represent object properties and solid lines represent subclass relations, e.g., *Host is-a Origin*.

2   TypOn – Typing Ontology

An ontology should make available both the vocabulary and the semantic rules required to proper represent knowledge of a given domain. In this section we provide an ontology suitable to describe knowledge in the microbiology typing methods domain, TypOn, depicted in Fig. 1. This ontology was developed and improved based on comments by domain experts and it constitutes a first proposal, that can be expanded and adapted as new typing methods are developed and already existing ones are updated. The ontology was developed with the help of the Protégé editor [11] and is available at http://www.phyloviz.net/typon.

The main aim of bacterial typing methods is the characterization of bacterial populations, where each sampled microorganism becomes an isolate, referring to the process of isolating it from the bacterial population. Thus, *Isolate* is a main concept for TyPon and it is characterized by several properties. An isolate belongs to a *Species*, property *belongsToSpecies*, which makes part of a *Genus*, property *hasSpecies*. The property *belongsToSpecies* has the property *hasIsolate* as its inverse. Moreover, for each *Isolate*, we know its *Origin*, either *Host* or *Environment*, its *GeographicInformation* and its *TypingInformation*. Note that a *Host* belongs also to a *Species* and that both *Host* and *Environment* may also have *GeographicInformation*. Although properties *hasGeographicInformation* and *hasOrigin* have usually cardinality at most one for each *Isolate*, the property *hasTypingInformation* has usually cardinality higher than one for each *Isolate*. For instance, an *Isolate* usually has available information for several typing methodologies such as MLST, antibiograms, etc. In this context, it is important to note that *TypingInformation* is the root of a class hierarchy which is extensible and that defines several typing methods (see Fig. 1). In particular, we are able to distinguish different categories of typing methods, *e.g.*, the ontology allow us to infer that *MSLTST* is a *Genotypic* technique and that, in contrast, *Antibiogram* is a *Phenotypic* technique.

As mentioned before, the current version of TyPon focus on MLST concepts, since it is the most widely used sequence based typing technique. In this context, we note in particular the concepts *Locus*, *Allele*, *MLSTSchema* and *MLSTST*. In MLST we can have several typing schemas described by a set of loci, each one being part of a sequence of an housekeeping gene. Such schemas are represented through the class *MLSTSchema*, which has the property *hasLocus*. Then, each *Isolate* may have associated one or more typing informations, obtained with different schemas, i.e., *MLSTST* instances, known as sequence types characterized by the observed alleles for each locus. Therefore, in our ontology, we associate to each *MLSTST* both a schema and the observed alleles through properties *hasSchema* and *hasAllele*, respectively. Notice also that *hasAllele* is a property shared by *MLSTST* and *Locus* classes and, thus, it does not have *isLocus* property as its inverse. It is also interesting to note that, by knowing only the *Locus*, it is possible to be aware of the *Species* that it belongs to, using the *isOfGene* and *belongsToSpecies* properties. The property *belongsToSpecies* is also an example of a property which has more than one class as domain.

We have also detailed the *Antibiogram* typing information technique in the current version. Namely, we have represented each *Antibiotic* as a concept, allowing the addition of new antibiotics as needed. The reaction of a given antibiotic is also represented as a concept, *AntibioticReaction*, allowing that each *Antibiogram* may have associated one or more antibiotic reactions, depending on the number of used antibiotics. These relations are given through the object properties *hasAntibioticReaction* and *hasAntibiotic*, respectively.

Additional information for each class, such as *id* and *other name*, are described through data properties. For instance, the class *GeographicInformation* has data properties such as *Country* and *Region*. The class *Isolate* has data properties such as *Strain* and *Year*.

3   RESTful Web services

A second contribution of our work is a RESTful API for making available microbial typing data represented through the above ontology. A Web services framework is under development, making use of the Jena Semantic Web Framework [9] and other standard Java technologies for developing Web services. The set of endpoints that were defined for retrieving microbial typing data include:

```
/services/typingmethods
/services/{typingmethod}
/services/{typingmethod}/genus
/services/{typingmethod}/{genusid}
/services/{typingmethod}/{genusid}/species
/services/{typingmethod}/{genusid}/{speciesid}
```

```
/services/{typingmethod}/{genusid}/{speciesid}/isolates
/services/{typingmethod}/{genusid}/{speciesid}/{isolateid}
/services/{typingmethod}/{genusid}/{speciesid}/sts
/services/{typingmethod}/{genusid}/{speciesid}/sts/{stid}
/services/{typingmethod}/{genusid}/{speciesid}/sts/{stid}/isolates
/services/{typingmethod}/{genusid}/{speciesid}/loci
/services/{typingmethod}/{genusid}/{speciesid}/loci/{locus}
/services/{typingmethod}/{genusid}/{speciesid}/loci/{locus}/{id}
```

The URI parameters, *i.e.*, the text inside {}'s, represent specific identifiers. For instance, {typingmethod}, {genusid} and {speciesid} should be parametrized with the name of the typing method (*e.g.* MLST), the name of the genus (*e.g. Streptococus*) and the name of the species (*e.g. pneumoniae*), respectively.

Each endpoint with {}'s at the end refers to a resource identified by a given id or unique label. As an example, with the endpoint

```
/services/{typingmethod}/{genusid}/{speciesid}/sts/{stid}
```

we may obtain the information of a specific sequence type. Moreover, with these kind of endpoints it also possible to replace their information, using the POST method. The other endpoints retrieve all individuals of a respective class. For instance, the endpoint

```
/services/{typingmethod}/{genusid}/{speciesid}/sts
```

retrieves all existing MLST sequence types in the database for the specified parameters {typingmethod}, {genusid} and {speciesid}.

We can also add more individuals with these kind of endpoints, using the PUT method. However, data deletion is only possible through the endpoints

```
/services/{typingmethod}/{genusid}/{speciesid}
/services/{typingmethod}/{genusid}/{speciesid}/{isolateid}
```

by using the DELETE method.

All endpoints return either text/html or application/json. There is also available a SPARQL endpoint and an authenticated endpoint to retrieve and submit data represented as rdf/xml. A more comprehensive description for the MLST data related endpoints is available at http://rest.phyloviz.net.

4   Implementation

A prototype Web client that makes use of the RESTful API and that allows users to explore and query data for some of the MLST public datasets, is also available at http://rest.phyloviz.net/webui/. In this prototype it is possible to query by MLST schema, *MLSTSchema*, by the id of the sequence type, *MLSTST*, and by locus, *Locus*. Also, the MLST schema and alleles can be downloaded in more than one format. A graphical visualization of isolate statistics is also available in this prototype.

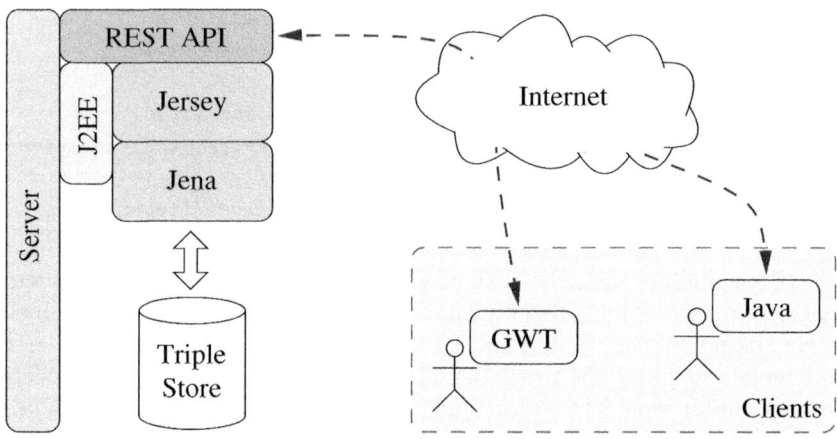

Fig. 2. Architecture of the Web service prototype. A REST API implemented over the Jersey framework, is made available, where data is accessed through the Jena framework. On the client side, we have implemented a Java REST client library and a Web application implemented over the Google Web Toolkit (GWT).

Our implementation makes use of the Jena Semantic Web Framework [9] and other standard Java technologies for developing Web services. Jena provides an API to deal with RDF data, namely a SPARQL processor for querying RDF data. In our implementation, both TypOn and all typing data are stored as RDF statements on a triple store. We are currently using the TDB triple store, a component of Jena for RDF storage and query. Although the Jena framework can use several reasoners, including OWL-DL reasoners, we are using the internal RDFS reasoner for validation purposes only. Nevertheless, given Jena flexibility, we can easily process our repository of statements through a more powerful reasoner, and insert inferred and relevant statements back to our repository. This is particularly useful whenever we update the ontology with new or equivalent concepts and properties, or when we want to index frequent SPARQL queries, in order to improve their speed. Moreover, under the open world assumption, with data distributed over several repositories, one may need to crawl and index several repositories, possibly instances of our Web service implementation, before proceed with reasoning and inference.

The REST API made available uses the Jersey implementation of JAX-RS (JSR 311), a Java API for RESTful Web services that provides support in creating Web services according to the REST architectural style. This implementation is an official part of Java EE 5 and it facilitates the implementations of RESTful Web services through the usage of annotations, simplifying the development and deployment of Web service clients and endpoints.

In the current implementation, any user can query the repository and only authenticated users can insert, update or delete data. A more refined authorization model is under development.

5   Final Remarks

The proposed ontology provides the basic concepts needed to establish the semantic relationships of the different sequence-based typing methodologies, and it is designed to allow further expansion. It should be easily expanded to encompass the newer NGS SNP typing techniques that are appearing in the microbial typing field, while providing a consistent link with legacy techniques and other databases. This Semantic Web approach for sharing microbial typing data also allows for local databases from different institutes and different methods to be connected through the use of specific REST endpoints.

Moreover, the proposed REST interface and ontology facilitates the decoupling between the information system and its possible client technologies, allowing the sharing of data in human- and machine-readable formats. This approach allows the design of novel interfaces between different databases and data analysis softwares, through the use of Web services mashups.

An immediate practical use of the framework is to provide the microbiology researchers with a quick and effective way to share data on new methods being developed based on sequence typing methods, since the creation of a new typing schema and adding its concepts on the ontology is straightforward. The information available for isolates typed using a new typing schema can then be parsed to RDF statements and uploaded to a server authenticated SPARQL endpoint and, then, a new database is automatically accessible. The GWT Web client provides then to the end-users a friendly interface for data access for querying and submitting new data.

Future work will focus on expanding the ontology and creating Web services to perform automated curation of data directly from sequencer files, in order to speed up the curation process, and ensure better quality and reproducibility of data in the field of microbial typing.

Acknowledgments. The work presented in this paper made use of data available at MLST.net [10], PubMLST [12] and Institut Pasteur MLST Databases [8].

References

1. Aguiar, S., Serrano, I., Pinto, F., Melo-Cristino, J., Ramirez, M.: Changes in Streptococcus pneumoniae serotypes causing invasive disease with non-universal vaccination coverage of the seven-valent conjugate vaccine. Clinical Microbiology and Infection 14(9), 835–843 (2008)
2. Fielding, R.: Architectural styles and the design of network-based software architectures. Ph.D. thesis, Citeseer (2000)
3. Harris, S., Feil, E., Holden, M., Quail, M., Nickerson, E., Chantratita, N., Gardete, S., Tavares, A., Day, N., Lindsay, J., et al.: Evolution of MRSA during hospital transmission and intercontinental spread. Science 327(5964), 469 (2010)
4. MacLean, D., Jones, J., Studholme, D.: Application of'next-generation'sequencing technologies to microbial genetics. Nature Reviews Microbiology 7(4), 287–296 (2009)

5. Mwangi, M., Wu, S., Zhou, Y., Sieradzki, K., De Lencastre, H., Richardson, P., Bruce, D., Rubin, E., Myers, E., Siggia, E., et al.: Tracking the in vivo evolution of multidrug resistance in Staphylococcus aureus by whole-genome sequencing. Proceedings of the National Academy of Sciences 104(22), 9451 (2007)
6. Spratt, B.: Multilocus sequence typing: molecular typing of bacterial pathogens in an era of rapid DNA sequencing and the internet. Current Opinion in Microbiology 2(3), 312–316 (1999)
7. Van Belkum, A., Struelens, M., De Visser, A., Verbrugh, H., Tibayrenc, M.: Role of genomic typing in taxonomy, evolutionary genetics, and microbial epidemiology. Clinical Microbiology Reviews 14(3), 547 (2001)
8. Institut Pasteur MLST Databases. Pasteur Institute, http://www.pasteur.fr/mlst/
9. Jena A Semantic Web Framework for Java. HP and Others, http://jena.sourceforge.net/
10. MLST: Multi Locus Sequence Typing. Imperial College of London, http://www.mlst.net
11. The Protégé Ontology Editor and Knowledge Acquisition System. Stanford Center for Biomedical Informatics Research, http://protege.stanford.edu
12. PubMLST. University of Oxford (UK), http://pubmlst.org/
13. Ridom SpaServer. Ridom bioinformatics, http://www.spaserver.ridom.de/
14. Streptococcus pyogenes emm sequence database. CDC, http://www.cdc.gov/ncidod/biotech/strep/M-ProteinGene_typing.htm

Multivariate Methods for the Integration and Visualization of Omics Data

Alex Sánchez[1,*], José Fernández-Real[2], Esteban Vegas[1], Francesc Carmona[1], Jacques Amar[3], Remy Burcelin[3], Matteo Serino[3], Francisco Tinahones[4], M. Carmen Ruíz de Villa[1], Antonio Minãrro[1], and Ferran Reverter[1]

[1] Departament d'Estadística, Universitat de Barcelona, Barcelona, Spain
[2] Institut d'Investigació Biomèdica de Girona, Girona, Spain
[3] Institut de Medecine Moleculaire de Rangueil, Toulouse, France
[4] Hospital Clínico Interuniversitario Virgen de Victoria, Malaga, Spain
asanchez@ub.edu

Abstract. As the developments in high throughput technologies have become more common and accessible it is becoming usual to take several distinct simultaneous approaches to study the same problem. In practice, this means that data of different types (expression, proteins, metabolites...) may be available for the same study, highlighting the need for methods and tools to analyze them in a combined way. In recent years there have been developed many methods that allow for the integrated analysis of different types of data. Corresponding to a certain tradition in bioinformatics many methodologies are rooted in machine learning such as bayesian networks, support vector machines or graph-based methods. In contrast with the high number of applications from these fields, another that seems to have contributed less to "omic" data integration is multivariate statistics, which has however a long tradition in being used to combine and visualize multidimensional data. In this work, we discuss the application of multivariate statistical approaches to integrate bio-molecular information by using multiple factorial analysis. The techniques are applied to a real unpublished data set consisting of three different data types: clinical variables, expression microarrays and DNA Gel Electrophoretic bands. We show how these statistical techniques can be used to perform reduction dimension and then visualize data of one type useful to explain those from other types. Whereas this is more or less straightforward when we deal with two types of data it turns to be more complicated when the goal is to visualize simultaneously more than two types. Comparison between the approaches shows that the information they provide is complementary suggesting their combined use yields more information than simply using one of them.

Keywords: Data Integration, Omic Data, Visualization, Multiple Factor Analysis.

---

* Corresponding author.

A.T. Freitas and A. Navarro (Eds.): JBI 2010, LNBI 6620, pp. 29–41, 2012.

1   Introduction

In this section we review the increasing need for data integration in biological research. In addition we discuss some approaches based on multivariate analysis to perform data integration.

1.1   Data and Technologies in Modern Molecular Biology

The post-genomic age, that is the first decade of the 21st century where human and other genomes have started to be available, has been characterized by the generation of huge quantities of biological data.

This has been facilitated not only by the increasing availability of genome sequences but also by the development of technologies that can be used, in a high throughput way to interrogate them by measuring many -probably hundreds or thousands- variables at the same time.

Examples of these approaches are DNA microarrays [13] that measure RNA as an indication of gene expression, protein chips that can be used to quantify the proteins in a sample [6] or SNP arrays [11] to measure single nucleotides polymorphism that is variation in genomic composition at individual positions. All these techniques can be characterized by the fact that they measure a single type of information -expression, concentration, allele etc.- on many variables -genes, proteins, position, etc.- simultaneously. In many cases the techniques can be described as belonging to one type of "omic" technology such as transcriptomics for the study of gene expression (the transcriptome), proteomics for the study of proteines (the proteome) and so on for metabolomics, lipidomics, etc. In a first epoch each of the previous approaches described above was adopted separately -mainly due to the fact that they were in initial stages and the technologies were unmature and often expensive. But as the decade went on the techniques became more stable and cheaper and it became obvious the interest of considering different types of data at the same time to gain a better understanding of the biological processes underlying the same problem.

As more data and possibilities to obtain them become available the interest of biological sciences is moving from componentwise study to trying to model and analyze organisms as a whole -in a systems biology approach. One cornerstone of this approach is obviously going to be the possibility of combining different data sources in a way that they contribute to a better understanding of the overall phenomena in study. This combination is the goal of data integration as considered here.

1.2   Data Integration: Its Meanings, Goals and Methods

The idea of data integration is not new and -what is worse- does not always have the same meaning.

The term is often used in a pure informatic way [8], to describe tools and methods to combine and interrogate multiple data sources such as Biological databases. This allows for instance that information about a topic which may be phisically disperse between different databases can be retrieved with a single query in such a way that the user doesn't realize its different origins. The protein resource Interpro (http://www.ebi.ac.uk/interpro/) which integrates more than a dozen protein databases is an example of this approach.

A different approach appears when one considers ways to combine related studies in order to obtain a consensuated stronger conclusion. Although this is not exactly data integration -it should rather be described as "results integration" or meta–analysis [1] much attention has been devoted in recent years to a mixed approach [14] which consists of combining studies –that is meta– analysis– by joining their data and re–processing and reanalyzing them again –that is data integration. Independently of how it is named one characteristic of this approaches is that they aim at increasing the power of previous studies by acquiring more data of the same type.

A third, widely adopted perspective, consists of considering different data types, measured or not on the same individuals and trying to combine them in such a way that they improve in some way the understanding of some of the underlying biological processes that have generated the data. This is the type of data integration that is the focus of this work. Adopting the notation in Hamid et al. [9] outlined in Figure 1, we are interested in integrating heterogeneous data types at a late stage that the goal of this work is to investigate how to use different data originated in different types of analysis and already preprocessed to make some type of integrative analysis that -in some way- increases the overall understanding of the problem being analyzed.

2   Integration Methodologies

High throughput technologies provide an unprecedented amount of data leading to new interpretation challenges in biology. Indeed, scientists are facing a lack of strategies to identify the genes and the gene products involved in different biological processes of interest. Despite the numerous efforts that have been made to tackle this problem it remains difficult to obtain from such data a concise visualization of the biological mechanisms involved in the situation under study.

In recent years there have been developed many methods that integrate the analysis of different types of data. Corresponding to a certain tradition in bioinformatics many methodologies are rooted in machine learning tools such as bayesian networks, support vector machines or graph-based methods. In contrast with the high number of applications from these fields, another that seems to have contributed less to "omic" data integration is multivariate statistics, which has however a long tradition in being used to combine and visualize multidimensional data. Given this relative bias we have investigated the application of multivariate statistics to integrate bio-molecular information.

What do we demand to a multivariate method to achieve biological knowledge from several sets of variables? In our opinion such method should be able to: 1) reduce dimension efficiently, 2) represent simultaneously samples and variables from each data set, 3) avoid the "small n, large p" problem and 4) integrate supplementary data in a common space joint with original data.

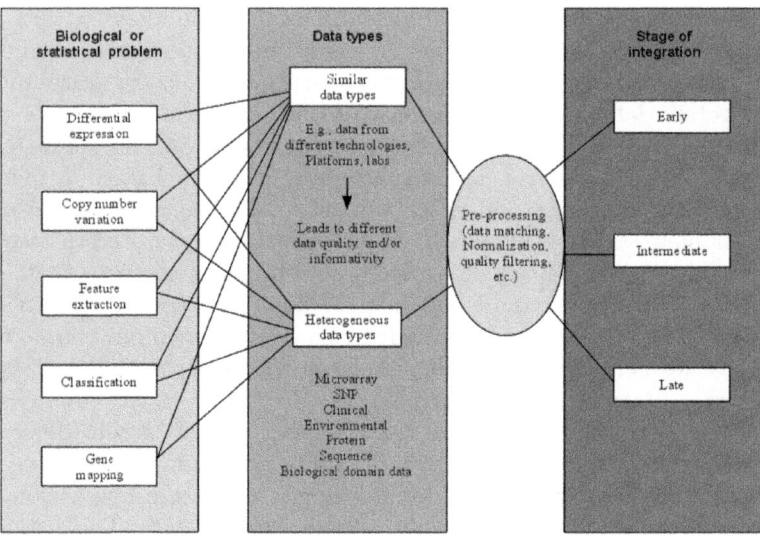

Fig. 1. The formulation of a data integration strategy has to account for three components: (i) which data are going to be integrated –that is what is the domain problem being analyzed-, (ii) the homogeneity of the data –that is are we going to combine homogeneous datasets, single datasets of different types of data or both– and (iii) At what stage is the integration going to be done –early preprocessing stages or late stages where data are ready for the analysis

Most of reference methods satisfies the first two points before mentioned. In contrast, last points are most difficult to satisfy.

The comparison between multivariate tecniques suited for this problem has been reviewed by [16] or [10]. Here we consider only Multiple Factor Analysis (MFA) in the sense of Escofier-Pagès [5],[4] which is appropriate because it verifies the requirements (1) to (4) stated above.

2.1   Multiple Factor Analysis

Multiple Factor Analysis (MFA) is dedicated to the simultaneous exploration of multi–way data sets where the same individuals are described by several groups of variables. MFA is commonly applied to sensory and ecology data and it has already been applied to the analysis of metabonomic data [3]. MFA has also been used to jointly analyze the structure emerging from genomic and transcriptomic data sets [2].

The assets of MFA appear when integrating both numerical and categorical groups of variables, and when supplementary groups of data need to be added in the analysis. Any of these operations can be performed in a natural way by MFA so the method is is particularly well adapted to integrate formalized biological knowledge.

The core of MFA is a PCA applied to the whole set of variables in which each group of variables is weighted, rendering possible the analysis of different points of view by taking them equally into account. Resulting graphical outputs for the first two principal components (PC1 and PC2) are used.

Multiple factor analysis (MFA) [5],[4] analyzes observations described by several "blocks" or sets of variables. MFA seeks the common structures present in all or some of these sets. MFA is performed in two steps. First, a principal component analysis (PCA) is performed on each data set which is then *normalized* by dividing all its elements by the square root of the first eigenvalue obtained from of its PCA. Second, the normalized data sets are merged to form a unique matrix and a global PCA is performed on this matrix. The individual data sets are then projected onto the global analysis to analyze communalities and discrepancies. MFA is used in very different domains such as sensory evaluation, economy, ecology, and chemistry.

The goal of MFA is to integrate different groups of variables by describing the same observations. In order to do so, the first step is to make these groups of variables comparable. Such a step is needed because the straightforward analysis obtained by concatenating all variables would be dominated by the group with the strongest structure. A similar problem can occur in a non-normalized PCA: without normalization, the structure is dominated by the variables with the largest variance. For PCA, the solution is to normalize (i.e., to use Z-scores) each variable by dividing it by its standard deviation. The solution proposed by MFA is similar: To compare groups of variables, each group is normalized by dividing all its elements by a quantity called its first singular value which is the matrix equivalent of the standard deviation. Practically, this step is implemented by performing a PCA on each group of variables. The first singular value is the square root of the first eigenvalue of the PCA. After normalization, the data tables are concatenated into a data table which is submitted to PCA.

Schematically MFA works as follows:

(I) The heart of MFA is a PCA in which weights are assigned to the variables:
   (i) When several sets of variables describe a same set of individuals, it is possible to consider the merged data set: $K = [K_1, K_2, \ldots, K_J]$, where each K_j corresponds to an "tipus" data table.
   (ii) Separate analysis are performed by principal components analysis (PCA) on each group j of variables. Each variable belonging to a group j is weighted by $1/\lambda_1^j$, where λ_1^j denotes the first eigenvalue of the matrix of variance-covariance associated with each data table K_j.
   (iii) A global analysis is performed. The corresponding graphical displays (Individual Factor Map and Variables Representation) are read as for PCA.
(II) MFA allows to look for common factors by providing a representation of each matrix of variables (Groups Representation). It provides the visualization of specific and common structure emerging from the K_j. MFA allows to compare the main factors of variability by linking both groups and variables representations. As the coordinates of set j upon axis of rank s is

equal to $L_g(z_s, K_j)$: set coordinates are always comprised between 0 and 1; and a small distance between two set along axis s means that they include the structure expressed by factor s each one with the same intensity.

(III) The asset of MFA to add supplementary groups of variables is used to integrate biological knowledge. The BP modules (groups of gens associated with a biological process) are formalized as matrices containing the restriction of the whole data set to the genes associated with the ith BP. The projection of the K_{BP_i} is made by means of its scalar product matrix between individuals. This matrix denoted W_i is a $(I \times I)$ matrix $(W_i = K_{BP_i} K'_{BP_i})$ and can be considered as an element of the space $\mathbb{R}^{I^2}$. This element is thus projected on the dimensions of issued from MFA. This representation of the groups is made available by means of a graphical display of the K_{BP_i} as points in a scatter plot. It has to be read as follow: the coordinate of a given group is all the more close to 1 than the variables of this group are highly correlated with the dimension issued from the MFA (either positively or negatively). Hence, two groups are all the more close than the structures they induce on the observations are close.

One of the interesting aspects of MFA is that it offers the possibility to integrate biological knowledge withe other variables. To do this each biological process is represented by a label (e.g. a GO term, call it "BP_i") which is assigned to a set of variables (eg genes). Thesw variables define a new matrix (K_{BP_i}). The K_{BP_i} matrices corresponding to all terms are included in the analysis as supplementary groups of variables that is they do not participate in the construction of the dimensions but they may be visualized along these dimensions. The interpretation of the results is made by means of the projection of the biological processes onto the dimensions issued from MFA. The more the coordinate of a biological process is close to 1, the more it is highly correlated with the corresponding dimension of variability between samples.

A paractical advantage for the use of MFA is the availability of a complete software [12] that can be used interactively or as an R package to perform all the analyses described in the paper.

3   The Problem for the Analysis

The problem used to illustrate the methods presented is the relation between obesity and insulin resistance [15]. Obesity –which is nowadays considered epidemics in some occidental countries– is a condition resulting from a chronic imbalance between energy intake and energy expenditure. In spite of what used to be common it does not result from character flaws but it has been proven to be a complex disease with molecular causes and highly heritable.

Obesity is associated with a metabolic problem known as *Insulin Resistance* which consists of cells becoming less sensitive to insulin, This provokes the pancreas to over-compensate by working harder and releasing even more insulin.

Combination of insulin-resistance and insulin over-production leads to two common outcomes: diabetes, or obesity combined with high cholesterol and increased rates of heart disease.

Being a complex disease Insulin Resistance can be explained by many different causes. The two main ones considered here are gene expression and the compoition of gut bacterial microflora (also called gut microbiota).

The data for this study consist of different types of variables all of which can be related with the study of obesity and insuline resistance. Essentially all the data can be associated with one of three groups: Insulin resistant (IR) Insuline Sensitive (IS) or Diabetic (Diab) individuals. Briefly these data types are:

1. Clinical variables which can be associated with obesity -such as weight, height, body mass index- or insuline resistance (Triglycerides, Cholesterol, Diabetes and other "usual" variables for this type of disease).
2. Denaturing Gradient Gel Electrophoresis (DNA bands) [7] is used to characterize microbial flora. These are discrete data, associated with electrophoretic bands each of wich can be associated with a different microbial profile for each group (IR, IS or Diab).
3. Expression microarrays [13] have become very common in biomedicine and given that obesity as well as insulin resistance are associated with changes in metabolic pathways so a reflection at expression level may be expected.

Alltogether faced with the study of this type of problem a biomedical researcher may have at least three data sets -each of which is separately informative- which she will be interested in combining in such a way that the information provided by each type can be reinforced by that provided by another type.

4   Results

MFA has been applied to analyze an unpublished data set about Insulin Resistance.

The resulting maps of samples and variables are presented in Figures 2, 3, 4, 5 and 6.

The Correlation circle plot (Figure 2) can be used to identify the variables most correlated with each principal component. This scatter plot representing the Expression, DGGE Bands and Clinic groups is read as in PCA. Briefly, the Expression, DGGE Bands and Clinical variables are projected on the factor map and represented as vectors. The more a vector has an absolute magnitude close to 1, the best the projection is. The vector points in the direction of the high values. We thus retrieved the genes most correlated with PC1 and PC2 (Figure 3). Table 1 shows genes whose correlation coefficient with PC1 or PC2 respectively has an absolute value greater than 0.85.

Once the identifiers recovered, they can be annotated manually by gathering functional information from a large panel of databases and annotation tools.

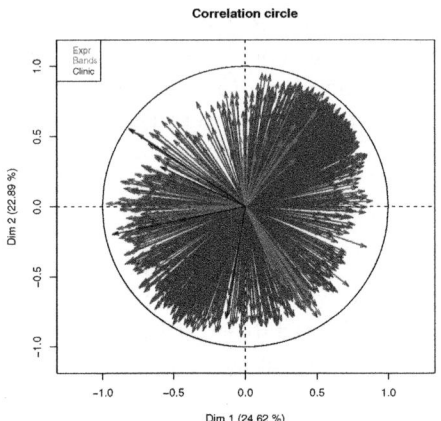

Fig. 2. The variables representation is displayed. The vector points in the direction of the high values.

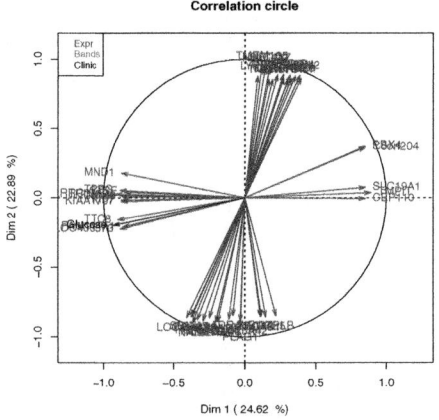

Fig. 3. Variables (genes) most strongly correlated with MFA factors can be identified. Genes with a correlation coefficient in absolute value greater than 0.85 are: C6orf204, CEP110 CNIH3, CTSE, FAM114A1, KIAA1797, KIF7, LOC400573, LRRC37A3, MND1, PBX4, SLC19A1, TMPIT, TPBG, TRIM6, TTC8,ADRA1B, AMD1, CHSY1, GGN, GNPNAT1, HECTD3, IL13RA1, KYNU, LDLRAP1, LOC116143, LOC401588, LYPD5, MDFI, NAT6, NETO2, NHEDC2, OAS1, PGPEP1,PLAG1, POLA2, POL-RMT, PTPLB , SEC14L5, SLC25A42, SLC4A7, SPATA7, tcag7.1196, TMEM165, TMEM2, TNFRSF6B, TNFRSF9 and XPNPEP3.

Table 1. Genes whose correlation coefficient with Principal Components 1 or 2 have absolute value greater than 0.85

Gene	Corr (PC1)	Gene	Corr (PC2)
C6orf204	0.853	ADRA1B	-0.851
CEP110	0.853	AMD1	0.859
CNIH3	-0.865	CHSY1	0.876
CTSE	-0.851	GGN	0.852
FAM114A1	-0.861	GNPNAT1	0.850
KIAA1797	-0.874	HECTD3	-0.889
KIF7	-0.873	IL13RA1	0.884
LOC400573	-0.882	KYNU	0.858
LRRC37A3	-0.912	LDLRAP1	-0.898
MND1	-0.871	LOC116143	-0.864
PBX4	0.853	LOC401588	0.860
SLC19A1	0.854	LYPD5	0.876
TMPIT	0.893	MDFI	-0.857
TPBG	-0.881	NAT6	-0.900
TRIM6	-0.972	NETO2	0.889
TTC8	-0.897	NHEDC2	0.937
		OAS1	-0.861
		PGPEP1	-0.881
		PLAG1	-0.933
		POLA2	-0.859
		POLRMT	-0.895
		PTPLB	-0.852
		SEC14L5	-0.855
		SLC25A42	-0.890
		SLC4A7	0.936
		SPATA7	-0.855
		tcag7.1196	-0.867
		TMEM165	0.948
		TMEM2	0.882
		TNFRSF6B	0.886
		TNFRSF9	0.851
		XPNPEP3	-0.875)

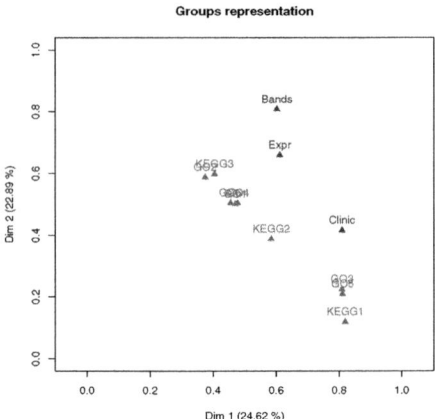

Fig. 4. Each group of variables is projected on the factor map created by MFA, active and supplementary groups are plotted.

As we pointed out, the interpretation of the structures emerging from MFA constitute a difficult and time-consuming step. Our approach aims at easing this task. Firstly, the biological knowledge is formalized. Here, Gene Ontology (GO) biological process (BP) terms and KEGG pathways are used to assemble gene modules. Secondly, the gene modules are superimposed on the same principal components and aid interpretation of the study (Figure 4). This is allowed by the capacity of MFA to integrate supplementary groups of data. The result has to be read as follow: the coordinate of a given group is all the more close to 1 than the variables of this group are highly correlated with the dimension issued from the MFA. Hence, two groups are all the more close than the structures they induce on the observations are close. Thus, the coordinate of one gene module provides a direct measure of the association between its constituting genes and the corresponding principal component. In Figure 4 we have plotted 9 supplementary groups of variables corresponding to most enriched KEGG pathways and GO categories. In particular, we represent 3 KEGG pathways: Id: 4060 (labeled as KEGG1), Id: 4062 (labeled as KEGG2) and Id: 4115 (labeled as KEGG3), and 6 GO terms: GO:0002376 (labeled as GO1), GO:0002460 (labeled as GO2), GO:0004871 (labeled as GO3), GO:0006950 (labeled as GO4), GO:0006954 (labeled as GO5) and GO:0060255 (labeled as GO6). We conclude that Clinic is more correlated with PC1 and Expression and Bands are more correlated with PC2. Also we observe that KEGG1, KEGG2, GO3 and GO5 are close to Clinic group and deduce that they induce a similar structure on the observations. On the other hand, GO1, GO2, GO4 and KEGG3 are more correlated with PC2 and are close to Expression.

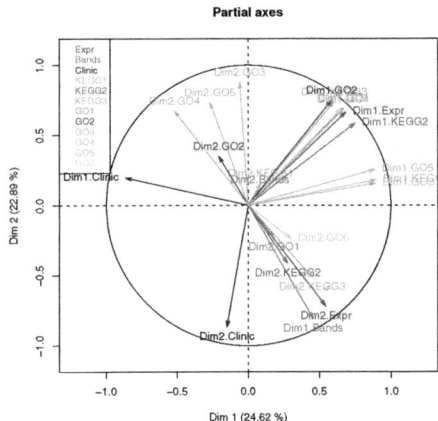

Fig. 5. Factors from separate analyses are represented with factors of MFA. KEGG1: Id 4060; Cytokine-cytokine receptor interaction. KEGG2: Id 4062; Chemokine signaling pathway. KEGG 3: Id 4115; Phagosome. GO1: Id GO:0002376; immune system process. GO2:Id GO:0002460; adaptive immune response based on somatic recombination of immune receptors built from immunoglobulin superfamily domains. GO3:Id GO:0004871; signal transducer activity. GO4: Id GO:0006950; response to stress. GO5: Id GO:0006954; inflammatory response. GO6: Id GO:0060255; regulation of macromolecule metabolic process.

Figure 5 is displayed to look at the link between the principal components of the MFA and the ones of each single group. Factors from separate analysis can be represented by the way of their correlations with factors of MFA. For the Expression group the first and second dimensions are similarly correlated with to the MFA's dimensions. For the DGGE Bands the first dimension is highly correlated with the MFA's second one, and the second dimension is highly correlated with the MFA's first one. For the Clinic group the first dimension is highly negative correlated with the first MFA dimension, and the second dimension is highly negative correlated with the second MFA dimension. We observe that the first dimension of the KEGG Bands is highly correlated with the second dimension of Expression. For the supplementary groups we observe that first dimension of each group are positively correlated with the first dimension of Expression group. KEGG1, GO3 and GO5 are highly correlated within them. Second dimension of KEGG2, KEGG3, GO1 and GO6 are positively correlated with the second dimension of MFA and the second dimension of KEGG1, GO2, GO3, GO4 and GO5 are negatively correlated with the dimension of MFA.

Figure 6 shows the representation of samples and it can be seen that MFA reduces the dimension efficiently. The structure of the samples appears nicely represented in three clusters. First MFA dimension separates Diabetic and Non diabetic samples, and second MFA separates insulin resistant and insulin sensible samples.

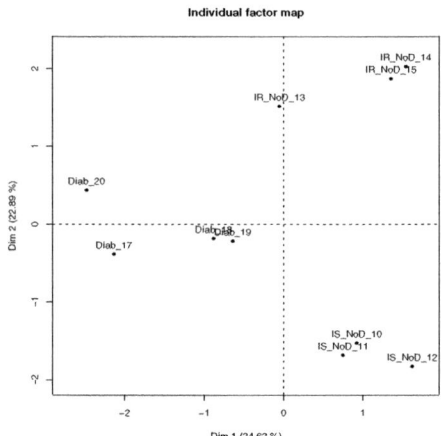

Fig. 6. Representation of samples

5   Discussion and Conclusions

When investigating complex diseases such as insulin resistance syndrome, it seems necessary to take into account all the informative experiments available. To tackle this challenging task, we propose to use MFA in such way that it becomes possible to combine data sets coming from different "omic" areas and to integrate biological knowledge with these data. MFA is firstly used to simultaneously analyze the structure emerging from the separate analysis of each molecular levels and to supply principal components which summarize parts of the data variability. The common structures are underlined and graphical outputs are provided such that biological meaning becomes retrievable.

MFA methodolgy is suitable for a wide range of biological investigations needing a comprehensive view of the datasets structures and an integration of their associated knowledge.

Acknowledgments. Alex Sanchez, Ferran Reverter and Esteban Vegas wisht to acknowledge the support of grant MTM2008-00642 of the Spanish Ministerio de Educacin y Ciencia.

References

1. Carlin, J., Normand, T.: Tutorial in biostatistics. meta-analysis: formulating, evaluating, combining, and reporting. Stat. Med. 19(5), 753–759 (2000)
2. de Tayrac, M., Lê, S., Aubry, M., Mosser, J., Husson, F.: Simultaneous analysis of distinct Omics data sets with integration of biological knowledge: Multiple Factor Analysis approach. BMC Genomics 10, 32–32 (2009)
3. Dumas, M., Canlet, C., Debrauwer, L., Martin, P., Paris, A.: Selection of biomarkers by a multivariate statistical processing of composite metabonomic data sets using multiple factor analysis. J. Proteome Res. 4, 1485–1492 (2005)

4. Escofier, B., Pages, J.: Analyses factorielles simples et multiples. [Multiple and Simple Factor Analysis], 3rd edn. Dunod, Paris (1998)
5. Escofier, E., Pages, J.: Multiple factor analysis (afmult package). Computational Statistics & Data Analysis 18, 121–140 (1994)
6. Falciani, F.: Microarray technology through applications. Taylor & Francis, New York (2007)
7. Gafan, G.P., Lucas, V.S., Roberts, G.J., Petrie, A., Wilson, M., Spratt, D.A.: Statistical analyses of complex denaturing gradient gel electrophoresis profiles. J. Clin. Microbiol. 43, 3971–3978 (2005)
8. Goble, C., Stevens, R.: State of the nation in data integration for bioinformatics. Journal of Biomedical Informatics 41(5), 687–693 (2008), http://dx.doi.org/10.1016/j.jbi.2008.01.008
9. Hamid, J., Hu, P., Roslin, V., Greenwood, C., Beyene, J.: Data integration in genetics and genomics: Methods and challenges. Human Genomics and Proteomics (2009)
10. Huopaniemi, I., Suvitaival, T., Nikkil, J., Orei, M., Kaski, S.: Multivariate multi-way analysis of multi-source data. Bioinformatics 26(12), i391–i398 (2010), http://bioinformatics.oxfordjournals.org/content/26/12/i391.abstract
11. Hao, K., Schadt, E.E., Storey, J.D.: Calibrating the performance of snp arrays for whole-genome association studies. PLoS Genet. 4(6), e1000109 (2008)
12. Lĕ, S., Josse, J., Husson, F.: Factominer: An r package for multivariate analysis. Journal of Statistical Software 25(1), 1–18 (2008), http://www.jstatsoft.org/v25/i01
13. Nguyen, D.V.: DNA microarray experiments: Biological and technological aspects. Biometrics 58(4), 701–717 (2002), http://www.blackwell-synergy.com/doi/abs/10.1111/j.0006-341X.2002.00701.x
14. Rhodes, D.R., Barrette, T.R., Rubin, M.A., Ghosh, D., Chinnaiyan, A.M.: Meta-analysis of microarrays: interstudy validation of gene expression profiles reveals pathway dysregulation in prostate cancer. Cancer Res. 62(15), 4427–4433 (2002)
15. Ricart, W., Fernández-Real, J.M.: Insulin resistance as a mechanism of adaptation during human evolution. Endocrinol Nutr. 57, 381–390 (2010)
16. Van Deun, K., Smilde, A., van der Werf, M., Kiers, H., Van Mechelen, I.: A structured overview of simultaneous component based data integration. BMC Bioinformatics 10(1), 246 (2009), http://www.biomedcentral.com/1471-2105/10/246

A Holistic Approach
for Integrating Genomic Variation Information

Pedro Lopes and José Luís Oliveira

DETI/IEETA, University of Aveiro,
Campus Universitário de Santiago,
3810 – 193 Aveiro, Portugal
{pedrolopes,jlo}@ua.pt

Abstract. Personalized medicine is strongly tied with human variome research: understanding the impact of specific genetic sequence mutations on observable human traits will play a key role in the quest for custom drugs therapies and improved patient care. Recent growth in this particular field leveraged the appearance of locus-specific databases (LSDBs). Although these systems are praised in the scientific community, they lack some features that can promote a more widespread usage. Existing systems are closed, independent and designed solely for gene curators. In this paper we present a new approach based on a holistic perspective of the genomic variation field, envisaging the integration of LSDBs, genes and variants, as well as a broad set of related resources in an innovative workspace. A prototype implementation for this approach is deployed online at http://bioinformatics.ua.pt/WAVe.

Keywords: LSDB, human variome, human genetics, bioinformatics, data integration.

1   Introduction

Personalized medicine has been the ultimate goal of multiple research projects over the last decade. Once the Human Genome Project [1] finished and the human genetic sequence was fully read, the perspective of designing custom drugs and preparing personalized treatments, fitting each patient's genetic characteristics, was the next logical step [2, 3]. However, despite tremendous achievements in this particular area, the goal of true personalized medicine and patient has not been attained yet.

There is a large gap between the current genetic sequence knowledge, the genotype, and the broad range of observable human traits, the phenotype. Hence, the genotype-to-phenotype research domain plays a key role in the establishment of associations between mutations in genes and medical conditions. For this matter, two complementary research lines are being adopted. On one hand, there are genome-wide association studies (GWAS). These association studies involve an *in silico* analysis of a large number of subjects, searching for statistical correlations between identified genotypes and respective phenotypes. On the other hand, there are genomic variation studies. This *in vitro* approach focuses on genomic mutations and their specific

A.T. Freitas and A. Navarro (Eds.): JBI 2010, LNBI 6620, pp. 42–49, 2012.
© Springer-Verlag Berlin Heidelberg 2012

impact on the human organism. Due to GWAS requirements regarding the huge number of samples, these studies have more noticeable results, thus generating more interest in the life sciences community. The genomic variation is currently growing steadily specially due to new variant description guidelines from the Human Genome Variation Society (HGVS) [4, 5] and new software packages, denominated locus-specific databases (LSDB). These tools were developed to manage the genomic variation ever-growing dataset. Although advances in this domain have fostered a quicker evolutionary pace, the poor overall quality and strict scope of current applications has hindered progress. LSDBs, detailed in the next section, are gene-centric, closed systems, designed for curators and fulfilling only the linear task of genomic variation storage and display. The diversity of concepts and resource types is innate to the human variome research filed, which leverages the need for more comprehensive and wide applications in this domain.

In this paper, a new holistic approach that attempts to overcome current deficiencies in the genomic variation software ecosystem is detailed. This approach is based on the enrichment of currently available datasets, connecting LSDBs, genes and variant information with multiple external resources and presenting them in an agile web-based workspace and through an easily accessible API.

2   Locus-Specific Databases

As soon as the genomic variation research field started evolving, researchers were faced with the cumbersome task of publishing their most relevant discoveries to the scientific community. In the beginning, many researchers simply listed the found variants in a non-coherent HTML table, Excel file or within publications. Moreover, there was a lack of quality control and scope in this area: available LSDBs are fragmented and closed systems, hampering the process of accessing and understanding available information. Modern developments leveraged the appearance of variation database management software packages such as LOVD [6], UMD [7] or MUTbase [8], thus reducing the effort required to find and browse discovered gene variants. MUTbase encompasses several web databases for advanced genomic variation studies with focus given to data organization and availability. UMD is very similar to MUTbase, however some of its locus-specific databases have private access. LOVD departs from UMD and MUTbase approaches by providing an "LSDB-in-a-box" perspective. Using LOVD, anyone can deploy its own locus-specific database complete with front- and back-office features as well as a list of services for programmable data access.

Current LSDBs, maintained by domain-specific experts, contain highly valuable information. The main outcome of this curation process is an extensive and curated genomic variation dataset, of critical relevance for researchers frequently in niche fields. Despite their undeniable quality, these systems possess some major flaws. Attempts to extract LSDBs true potential usually hit the interoperability drawback: they are closed and independent systems. Therefore, data aggregation or integration within distinct systems is impossible to achieve. Hence, exchanging data between distinct LSDBs or between LSDBs and other applications is not trivial. Furthermore, LSDBs are targeted at gene curators, which results in a focused and limited set of data related to the gene in hands.

Nowadays, projects like PhenCode (http://globin.bx.psu.edu/phencode), Genotype-to-Phenotype: A Holistic Solution (http://www.gen2phen.org) or MutaDATABASE (http://www.mutadatabase.org) are attempting to improve these platforms and other genotype-phenotype database applications [9]. Ongoing research focus includes standardizing LSDB structure and normalizing formats for data exchanges between heterogeneous systems. Nevertheless, legacy systems not adhering to new standards are being withdrawn and their role in future discoveries is compromised.

A need for distinct software tools, capable of tackling the combination of problems in both modern and legacy LSDBs, arises. Researchers require access to miscellaneous resources and features during their experiments workflow: a new approach must be adopted, accommodating not only gene and variant information, but also connections to external resources like proteins, diseases, publications or drugs. Furthermore, the aggregation of all available LSDBs (and their datasets) in a single central system should also be accomplished, providing a general vision over the entire genomic variation landscape.

3   An LSDB Aggregation Web System

3.1   Requirements

Division, and consequent diversity, in the LSDB ecosystem results in an extremely heterogeneous and disjointed mesh of independent data silos, each adopting its own standards and approaches. This lack of harmonization harnesses interoperability and integration, making data aggregation impossible in current circumstances. The development of a new LSDB aggregation system must take in account these domain specificities, modern application trends, and user expectations regarding available information and features.

The key requirement revolves around the aggregation of publicly available genomic variation datasets from miscellaneous and distributed locus-specific databases. These should be enriched with connections to external resources, thus broadening the scope of the available information displayed to researchers.

Next, integration and interoperability among LSDBs and between LSDBs and other systems is also important. Not only should new systems collect information, but also permit the inclusion of these data in other external systems. Moreover, the life sciences research field is dynamic: data hierarchical organization changes are frequent and new resources also surface constantly. Consequently, integration and interoperability should be complemented with extensibility. Regarding data integration, new systems should also enrich genomic variation datasets with connections to external resources, to widen available information scope. At last, content authorship and ownership should be maintained. That is, original integrated application should be linked and not hidden. Fulfilling these complex requirements triggered the design of a lightweight integration architecture, where integration and extensibility are key facets.

3.2   Architecture

The outlined strategy to achieve the proposed goals revolves around the idea of a single scalable schema providing a holistic view on the genome variation field. For this, it should be composed of a core, genes and associated variants, which can be easily augmented with extensions, such as connections to proteins or diseases for instance (Fig. 1).

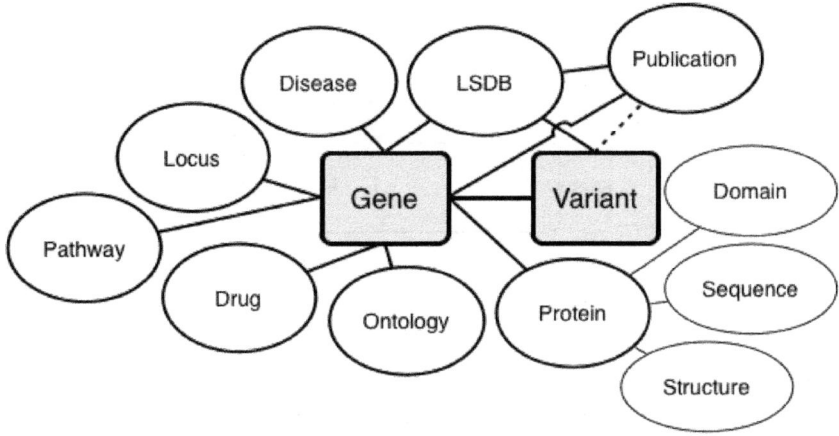

Fig. 1. Proposed holistic approach: gene and variant datasets are connected to multiple distinct resources, creating a rich and unique information network

This model relies on a lightweight link-based data integration architecture [10] to connect multiple distinct resources. With this approach, new applications can cover a wider range of data types, overcoming scope limitations that define currently available systems.

This implies that only resource identifiers are stored, and they will act as pointers to the original content when the users are browsing the application. By combining this approach with the "core plus extensions" model, developers can easily add new resources, of any kind, to the original package. Adopting the presented model to a feasible platform, while taking in account application scalability and requirements range resulted in a complex architecture (Fig. 2), involving interactions amongst five central components: configuration file, build engine, resource connectors, database and client applications.

The configuration file is written in XML and is structured in two distinct sections: a static module, for storing settings regarding the core data, genes and variants; and a dynamic module, where the miscellaneous extensions, and respective resources, are configured.

To further improve the platform's scalability, distinct connectors were deployed, allowing collecting extensions data within the same working environment. Resources for extensions can be gathered using CSV, XML, SQL or Web Service formats, each using its own settings. Data gathering tasks are executed by the platform's build engine: it reads settings from the configuration file and loads data to the database.

Once the database is fully populated it is the main data source for diverse client applications. Along with a web platform using direct access to data, an API must also be made available so that bioinformatics developers can use collected information within existing or new applications.

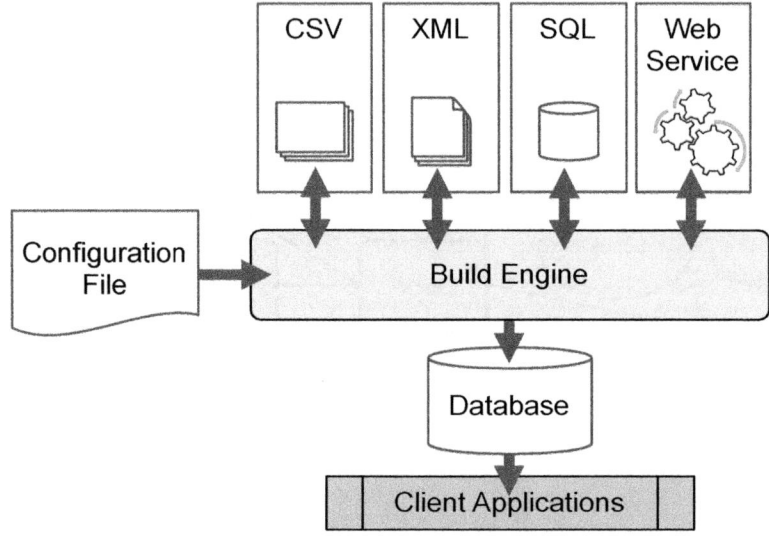

Fig. 2. Prototype architecture for proposed approach implementation, including settings file (*Configuration File*), connectors (*CSV, XML, SQL, Web Service*), build engine (*Build Engine*), database *(Database)* and client application (*Client Applications*)

By implementing this architecture, a real-world application would be able to present a holistic view over genomic variation information, encompassing data from miscellaneous resources, in a central working environment. This would include connecting distributed and heterogeneous LSDBs, aggregating genomic variation datasets and enriching the latter with the most relevant gene-related information.

3.3   Result

To assess the feasibility of the proposed approach, a prototype application was setup and made available for user testing. The prototype components were developed using Java and the data model was streamlined to a MySQL database. This new application, Web Analysis of the Variome (WAVe), is available online for public usage at http://bioinformatics.ua.pt/WAVe.

WAVe's entry page displays a gene search box and accessing a gene's workspace is as easy as typing the gene HGNC-approved symbol: this will trigger the automatic suggestions mechanism, leading directly to the selected gene workspace interface. Querying for "*" lists all genes with their respective LSDB and variant count. A checkbox at the top of the browse interface enables filtering for genes that contain at least one locus-specific database. This way, users can focus on expertly curated genes during their search.

The gene workspace interface is organized in two main sections: the navigation sidebar at the left, and the main content window at the right (Fig. 3). The complexity inherent to data navigation following the proposed holistic approach was hidden behind a simple and straightforward tree. Starting from the gene, users can browse all available resources by traversing the tree's nodes. Node leafs represent direct pointers to external resources and trigger the "LiveView" mode, where connected resources are loaded in the content window, within WAVe. This enables direct access to original data without leaving the application and, consequently, without breaking the research workflow. The sidebar also contains a "LiveView" toolbox, enabling users to expand or collapse the "LiveView" window size, open "LiveView" in a new window, return to the gene information summary page or access a gene feed. The latter allows direct access to the system API, where all gene-related content is available to any feed reader or programming language.

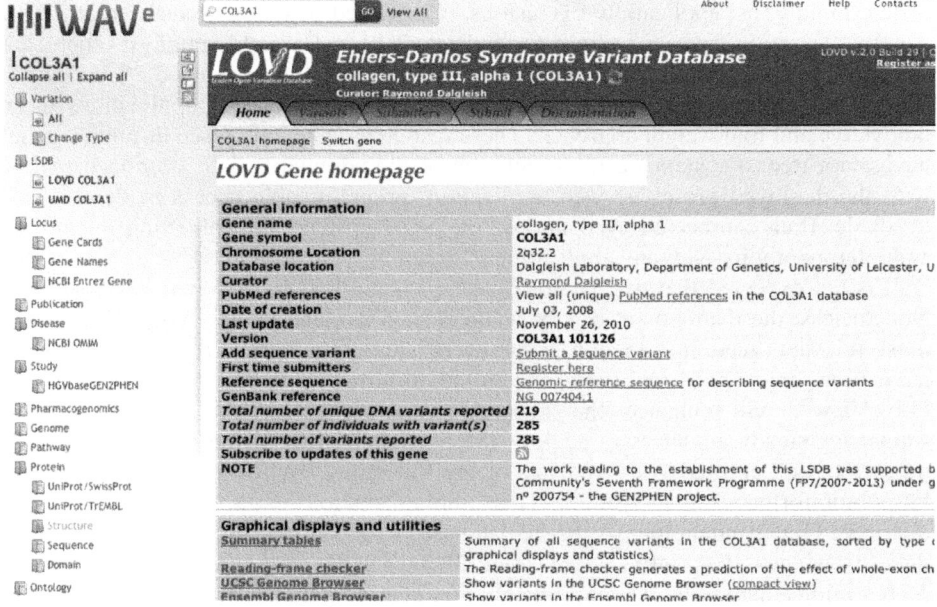

Fig. 3. Web Analysis of the Variome (WAVe) gene workspace for COL3A1. Search and main menu (*top*); gene navigation tree and "LiveView" toolbox (*left*); and COL3A1 LOVD instance in "LiveView" (*centre right*)

In addition to the deployed web portal, data can also be accessed through the previously mentioned API. Data for the human COL3A1 gene can be retrieved in feed format (Atom or RSS) from http://bioinformatics.ua.pt/WAVe/gene/COL3A1/atom or in JSON format from http://bioinformatics.ua.pt/WAVe/gene/COL3A1/json.

Browsing the various available resources, researchers can access directly to miscellaneous applications. The wide, and growing, amount of available data includes pointers to UniProt, Ensembl, NCBI, GeneCards, HGVbaseG2P, HGNC, Expasy, InterPro, PDB, PharmGKB or KEGG, among others. Current platform build contains locus-specific information for around 10% of known genes, approximately 86 thousand variants and more than 1 million pointers.

4   Conclusion

The presented approach introduces an innovative strategy for genomic variation data integration. Unlike traditional LSDBs, applications using this approach will provide a wider range of features and resources, resulting in a more generic system. The emergence of these systems will result in an improved application landscape, where both genomic experts and physicians are able to obtain deeper insights from a particular set of variation data.

A prototype tool was developed, as proof-of-concept using the detailed approached for genomic variation integration and data integration without harming content authorship or ownership. Moreover, any kind of online-available resources can be added rapidly as application extensions. Connected resources, such as LSDBs, genomic variation datasets or links to UniProt, PDB or Ensembl, amongst others, are directly available within the application, through a feature named "LiveView". "LiveView" allows contextual navigation between heterogeneous applications in an innovative and lightweight approach. These resources can be browsed through a gene-navigation tree: a systematic representation of the adopted holistic approach. Along with the deployed prototype web application, programmable access to data is also available. Data can be retrieved in feed or JSON formats, thus allowing integration and interoperability with any application.

It is not expected that true personalized medicine will be achieved in short term. Nonetheless, the deployment of key features such as gene holistic vision, where the most relevant gene information is at users' fingertips through a straightforward navigation tree, and real-time access to original external resources through "LiveView", will enhance the genomic variation researchers' workflow, thus supporting steady advances to achieve the personalized medicine panacea. The new prototype tool, entitled Web Analysis of the Variome (WAVe), is freely available at http://bioinformatics.ua.pt/WAVe. This implementation extracts true value from distributed independent locus-specific databases by means of indirect connections and data exchanges. Currently, more than 3800 LSDBs are connected in WAVe, representing a dataset with approximately 86 thousand genomic variants.

Acknowledgments. EC (FP7/2007-2013) under grant agreement 200754 (project GEN2PHEN) funded this work.

References

1. Watson, J.D.: The human genome project: past, present, and future. Science 248, 44–49 (1990)
2. Bottinger, E.P.: Foundations, promises and uncertainties of personalized medicine. Mount Sinai Journal of Medicine: A Journal of Translational and Personalized Medicine 74, 15–21 (2007)

3. Ginsburg, G.S., McCarthy, J.J.: Personalized medicine: revolutionizing drug discovery and patient care. Trends in Biotechnology 19, 491–496 (2001)
4. Ring, H.Z., Kwok, P.-Y., Cotton, R.G.: Human Variome Project: an international collaboration to catalogue human genetic variation. Pharmacogenomics 7, 969–972 (2006)
5. Cotton, R.G.H.: Recommendations of the 2006 Human Variome Project meeting. Nature Genetics 39, 433–436 (2007)
6. Fokkema, I.F.A.C., den Dunnen, J.T., Taschner, P.E.M.: LOVD: Easy creation of a locus-specific sequence variation database using an ldquoLSDB-in-a-boxrdquo approach. Human Mutation 26, 63–68 (2005)
7. Béroud, C., Collod-Béroud, G., Boileau, C., Soussi, T., Junien, C.: UMD (Universal Mutation Database): A generic software to build and analyze locus-specific databases. Human Mutation 15, 86–94 (2000)
8. Riikonen, P., Vihinen, M.: MUTbase: maintenance and analysis of distributed mutation databases. Bioinformatics 15, 852–859 (1999)
9. Thorisson, G.A., Muilu, J., Brookes, A.J.: Genotype-phenotype databases: challenges and solutions for the post-genomic era. Nat. Rev. Genet. 10, 9–18 (2009)
10. Lopes, P., Arrais, J., Oliveira, J.L.: Link Integrator - A Link-based Data Integration Architecture. In: Fred, A.L.N. (ed.) International Conference on Knowledge Discovery and Information Retrieval, pp. 274–277. INSTICC Press, Funchal - Madeira (2009)

SB-KOM: Integration of Pathway Information with BioPax

María Jesús García-Godoy, Ismael Navas-Delgado, and José F. Aldana-Montes

Computer Languages and Computing Science Department,
University of Malaga, Malaga 29071, Spain
{mjggodoy,ismael,jfam}@lcc.uma.es

Abstract. BioPax Level 3 is a novel approach to describe pathways at a semantic level by means of an owl ontology. Data provided as BioPax instances is distributed in several databases, and so it is difficult to find integrated information as instances of this ontology. Biopax is a biology ontology that aims to facilitate the integration and exchanged data maintained in biological pathways data. In this paper we present an approach to integrate pathway information by means of an ontology-based mediator (SB-KOM). This mediator has been enabled to produce instances of BioPax Level 3 from integrated data. Thus, it is possible to obtain information about a specific pathway extracting data from distributed databases.

Keywords: Biopax Level 3, AMMO, Pathways, SB-KOM, Ontology, Semantic Web.

1   Introduction

For several years now there has been an exponential growth of the amount of life science data obtained by biochemical and biologic experiments. Large amount of biological data are distributed in the Web and the degree of semantic heterogeneity is very variable so we need that all of this information must be combined and interconnected with other information by different bioinformatics tools to obtain new information from similarities, occurrences or patterns that it cannot be observable easily by scientists.

Integrated data from different biochemical and molecular biology areas can optimize investigation of biologists. In this context, many bioinformatics groups are developing tools that can help to integrate biological information in different levels [1]. One of these tools is Biopax (http://www.biopax.org/index.php). Biopax is an ontology in OWL format that aims to facilitate the integration and exchange of data maintained in biological pathway databases. Biopax can support the data model of a number of pathway databases such as ByoCyc [2], BIND [3], PATIKA [4], Reactome [5], aMAZE [6], KEGG [7], INOH [8], NCI/Nature PID [9], PANTHERS Pathway [10] and others. Biopax has been designed in three levels, each level was designed to specific type of pathways: Biopax Level 1 was designed to represent metabolic pathway data; Biopax Level 2 includes representation of molecular binding interactions and hierarchical

A.T. Freitas and A. Navarro (Eds.): JBI 2010, LNBI 6620, pp. 50–65, 2012.
© Springer-Verlag Berlin Heidelberg 2012

pathways; and Biopax Level 3 adds support for representation of signal transduction pathway, gene regulatory networks and genetic interactions. There are several databases providing their data in Biopax format. This enables database interoperability but these data are not usually integrated.

On the other hand, AMMO (http://www.asp.uma.es) is an ontology [11] that has been used to integrate metabolic information by means of an ontology-based mediator (SB-KOM) [12]. This mediator was used to integrate the most relevant public databases at different levels of study that includes metabolites properties, concentrations, macromolecular structures, assigned functions, docking among macromolecules and information about biochemical pathways. These integreted data have been used to produce a graphical tool, the System Biology Metabolic Assistant [13] (SBMM), thar was designed to retrieve metabolic information, to build metabolic models and to annotate them in SBML.

However, AMMO is not a standard for pathway representation, and this limits its use to SBMM tool. In this paper we face the production of BioPax instances from integrated data using SB-KOM. In this sense, we have to define the relationships between datasources and this onotlogy. In order to take advantage of previous knowledge, we have firstly made a comparative study between AMMO and Biopax Level 3 to determine the differences and similarities between them.

The paper is organized as follows: Section 2 describes SB-KOM mediator; Section 3 shows how SB-KOM can be configured to access pathway data using BioPax Level 3; Section 4 introduces some use cases; Finally we conclude with some remarks in Section 5.

2   Previous Work

KOMF, the Khaos Ontology-base Mediator Framework [14] (available at http://www.khaos.uma.es/KOMF) enables the implementation of the internal components of a mediation system in which the domain of knowledge is represented by means of OWL domain ontologies. The primary goal of KOMF is to integrate databases, which are accessible via Internet or can be downloaded for local use. This mediator enables a way to produce integrated data.

The KOMF architecture follows an ontology-based approach, which enables the reuse, especially in life science, of the available ontologies for data integration. Besides, KOMF components are functionally independent, so developers can produce integration systems without implementation complexities and can have the possibility of extending them by including new components. In addition, wrappers are published as Web Services, which allows them to be easily reused, by other applications and mediators. The mappings between data source schemas and different domain ontologies can be defined by means of a metadata repository. This gives the ability to applications to use different ontological concepts when querying integrated data sources. The configuration needed for integrating data is limited to the definition of a set of mappings between domain ontologies and the Data Service Schemas.

This framework has been instantiated, producing the instantiated mediator SB-KOM, for integrating molecular biology data sources from different biological information sources related to Biochemistry, Molecular Biology and Physiopathology. In this context, it is being used in real systems like the System Biology Metabolic Assistant (SBMM), where different databases have been integrated, such as Brenda [15], KEGG, CheBI [16], Uniprot [17] and Sabio-rk [18].

3   SB-KOM Configuration

The integration of data using SB-KOM requires the configuration of the mediator by: registering the ontology to be used (BioPax Level 3 in this case), defining the mappings between the domain ontology and the data sources and finally registering those mappings (Figure 1).

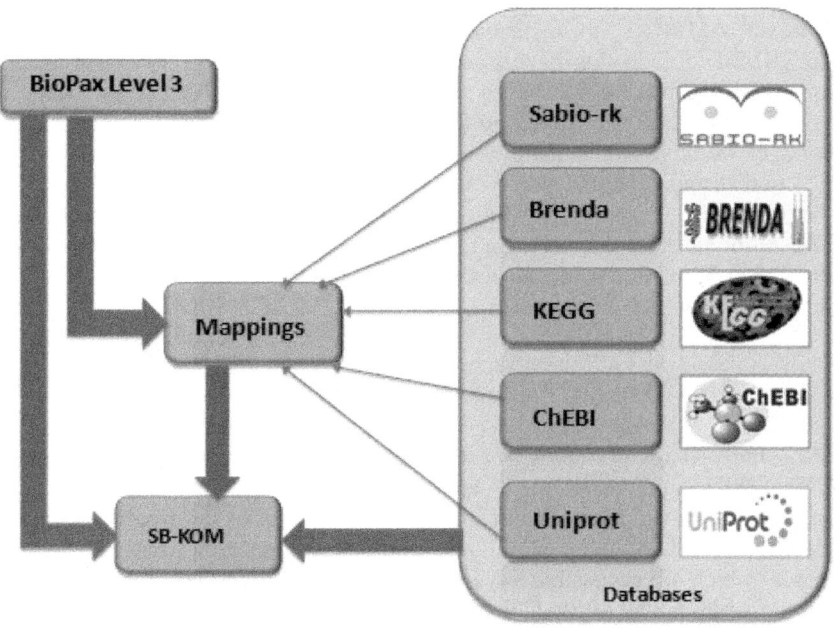

Fig. 1. Configuration and use of SB-KOM mediator

As the main part for integrating data using SB-KOM is the definition of mappings, which we could face using two different strategies:

1. To analyze the different data source schemas and try to align them to BioPax Level 3. This is a costly and slowly process that requires testing that each mapping works properly.
2. To take advantage of the mappings already defined between AMMO and the data source schemas, and try to align AMMO and BioPax Level 3 to translate those mappings to SB-KOM. In this case the mappings defined between AMMO and the data sources has been tested and have demonstrated their utility, so only few new mappings need to be checked.

Thus, the methodology used in this approach to obtain the corresponding mappings is divided in three parts: 1) Analyzing the structure of Biopax Level 3 ontology; 2) Detecting differences and similarities of classes and properties between AMMO and Biopax Level 3; 3) Translating mappings of AMMO to mappings with Biopax Level 3.

The result of steps 1 and 2 is a list of mappings (relationships) between AMMO and BioPax Level 3 ontologies. In Table 1 we show the concepts that are similar between these two ontologies, and the definitions used to determine these similarities.

We have determined correspondences between classes, subclasses and properties in AMMO and Biopax Level 3. Some examples of the mappings between the ontologies presented in Table 1 are:

− Amino-acid sequence class in AMMO corresponds to Protein subclass in Biopax Level 3;
− Cellular organism class in Biopax Level 3 refers to Organism property in AMMO;
− Source Tissue class in AMMO corresponds to Tissue property in Biopax Level 3.

On the other hand, no correspondences have been determined; the sequence subclass of Amino acid sequence class and Deoxiribonucleic acid class in AMMO hasnot any correspondences in Biopax Level 3 ontology; SBML class, a class refers to a SBML format in AMMO hasnot any correspondence in Biopax Level 3.

Using the mappings between AMMO and BioPax Level 3, we have determined the mappings between BioPax Level 3 and the data sources available to be integrated in SB-KOM. These mappings are shown in the Appendix. This enables us to retrieve integrated data using SB-KOM and the BioPax terminology. Thus, results obtained will be integrated instances of this ontology.

4   Use Cases

Based on the mappings (see Appendix) obtained from results shown in Table 1, we have performed four use cases to test the functionality of SB-KOM using BioPax Level 3 as the domain ontology. In these use cases we aim to extract as many information as possible for each pathway. The pathways chosen for these use cases are: glycolysis/gluconeogenesis, fatty acid biosynthesis, pentose phosphate pathway and Krebs cycle.

In order to retrieve information about these pathways we have developed a set of queries that are sent to SB-KOM. These queries are:

```
1. Ans(P):-Protein(P),displayName (P, "lanosterol synthase")
2. Ans(P):-Pathway(P),displayName (P, "pathwayID")
3. Ans(M):-Smallmolecule(M), standardName(M, "Lanosterol")
4. Ans(R):-BiochemicalReaction(R), displayName(R, "R70485")
```

Table 1. Mappings of AMMO and BioPax Level 3 concepts

AMMO	Biopax3	Description Biopax3
Aminoacidic sequence class	Phisical entity class. Protein subclass and Small molecule subclass.	Physical entity presents an aminoacidic sequence
Cellular organism class	Organism property	The property Organism has a domain that is composed by Pathway, reference DNA, reference RNA, reference protein, gene, region of DNA and RNA reference; the range is Biosource
Correlationated reactions set class	Pathway class. Biochemical reaction subclass (Conversion class)	Pathway class refers to an interactions set. Biochemical reaction class is a sub-set of Correlationated reactions set class. It can be that Biochemical Reaction sub-class of the conversion class of Biopax3, subset of Correlationated reactions set in AMMO, property "is a collection of"
Sequence of desoxyribonucleic acid class	Physical entity has DNA and DNA region subclass	DNA class refers to a polymer whose monomers are nucleotide (ribose + nitrogen base + phosphate).
Experimental conditions class	Delta G property and Delta G subclass of Utility class. Kprime class and Keq subclass	In Biopax3, Delta G subclass presents Temperature and PH properties K´ subclass has Temperature and PH properties
External bibliography referente class	Xref subclass (subclass of the Utility class)	In Biopax3, Xref (subclass of the Utility class) is defined as the reference to an external source of information to a determinate instance, it corresponds to External bibliography reference class in AMMO
Functional compartment class	Cellular location property. Property of the class Physical entity and Cellular location vocabulary	In Biopax 3, the property Cellular location refers to a place in the cell (example: the cytoplasm). This property of the Physical entity class and Cellular location vocabulary coincides with the Functional compartment of AMMO ontology
Mathematics expression class	No correspondence of class. No correspondence of Biopax3 properties	In Biopax 3, there isn´t any class that includes Mathematics expression. However, there are subclass of Utility class as K prima and Delta of G that can be considered subsets of Mathematics expression in AMMO
3D structure class	Structure property in Biopax3. Chemical structure class in Biopax3.	Chemical structure class in Biopax3 refers to 3D structure of small molecules. Chemical structure class in Biopax3 could be considered as a subset of Chemical structure in AMMO, this could include macromolecules and not only small molecules.
Molecular Entity class		In Biopax3, Entity class presents Physical entity, inside it can distinguish: Complex, DNA, DNA region, Protein (example, receptor for insulin), RNA region, RNA, Small molecules (example aeroplisinin-1)
Molecular entity class includes two subclass: Compound and Macromolecule	Physical entity subclass of the Entity class	In Biopax3, products, substrates and modifiers can be considered a subset of the small molecules class and enzymes can be considered a subset of the Protein class

Table 1. (*continued*)

Nucleic acid sequence class	DNA, DNA region, RNA, RNA region (Subclass of physical entity class)	In Biopax3, physical entity class includes some subclass: DNA, DNA region, RNA and RNA region corresponds to Nucleic acid sequence class in AMMO. Gene class in Biopax 3 can be considered nucleic acid
Participant class	Entity class includes Gene, Interaction, Pathway and Physical entity subclass. Physical entity is constituted by Complex, DNA, DNA region, Protein, RNA, RNA region and Small molecule subclass	Entity class includes Gene, Interaction, Pathway and Physical entity subclass (and others) : Gene subclass refers to a location in the genome that transcribes to RNA and then, it has the capacity to affect "cellular event" Physical entity class is formed by Complex, DNA, Protein, RNA, Region of RNA and Small molecule subclass
Pathway class	Pathway subclass of Entity class	In Biopax3, Pathway subclass refers to a set/series of interactions that often form a network It doesn't correspond to Step of pathway class because it refers to a only one reaction at same in Biochemical reaction subclass (Subset of Step of pathway)
Quantitative parameter class This class includes biochemical parameters and concentrations subclass	Class could be considered a subset of the Quantitative parameter class	In Biopax3, K class is a biochemical parameter, subset of Biochemical parameter in AMMO. Delta G class is Gibbs's free energy, subset of Biochemical parameter subclass in AMMO.
Reaction class	Biochemical reaction subclass Subclass of Conversion class	Biochemical reaction refers to an interaction of conversion in which one or more entities (reactives) are transformed to an entity or more (products) by covalent changes On the other hand, Chemical reaction class has a property (and others) that is the participant property that includes the controller (property of catalysis, enzyme) and elements that constitute the chemical reaction
Ribonucleic acid class	Physical entity class includes RNA y Region of RNA subclass	In Biopax 3, Physical entity class includes RNA, RNA region subclass (and others). This subclass describes polymers or regions of polymers that their monomers are constituted by ribose, nitrogen base and phosphate group
Tissue class	Tissue class in AMMO corresponds to tissue property in Biopax3	Biosource subclass (subclass of Utility class) refers to the origin source of macromolecules or compounds that can be tissues (entity) Tissue Vocabulary subclass refers to BRENDA (ontology de tissue) Tissue property of Biopax3 corresponds to issue class in AMMO. In Biopax, Pathway subclass refers to a set of connecting entities and a tissue is a set of interconnecting cells.

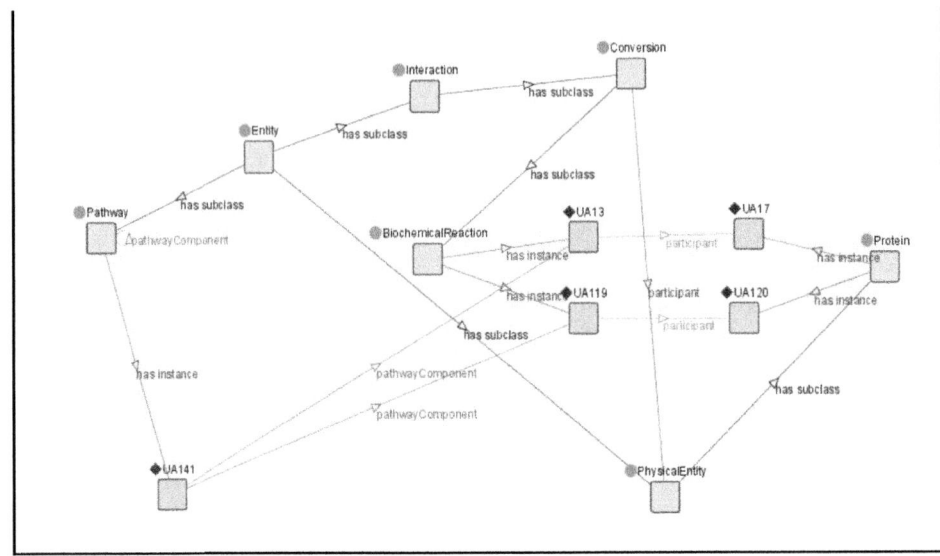

Fig. 2. Partial view of glycolysis/gluconeogenesis pathway. Yellow square nodes represent classes and grey square nodes represent instances. The lines of different colours represent the relationship between elements (classes, subclasses and instances).

Figure 2 and Figure 3 represent a partial view of the glycolysis/gluconeogenesis and fatty acid biosynthesis respectively extracted as instances of BioPax Level 3. Figure 4 shows classes and instances of glycolysis pathway and relationships (properties) between them. Figure 3 represents a detailed view of instances of fatty acid pathway in Biochemica reaction class that includes all biochemical reactions of the pathway.

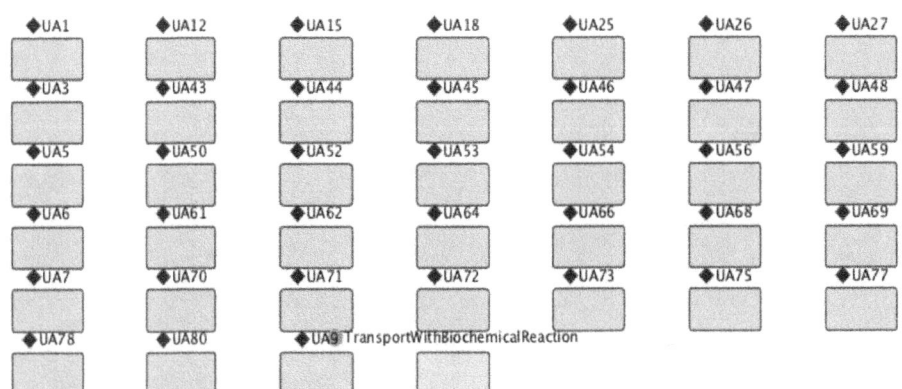

Fig. 3. Partial view of fatty acid biosynthesis. Yellow square nodes represent classes and grey square nodes represent instances. The lines of different colours represent the relationship between elements (classes, subclasses and instances).

Figure 4 represents the density instances and class of glycolysis pathway and it shows instances and classes in a separated view and the relationships between instances and classes independently.

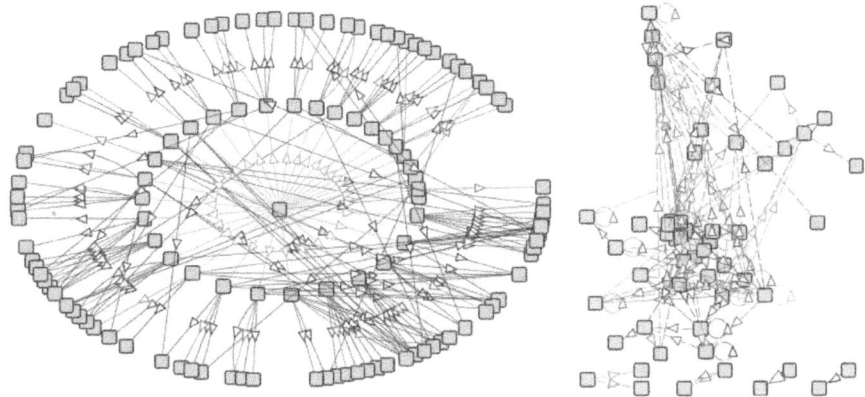

Fig. 4. Density of instances represented in grey and subclases represented in yellow in glycolysis pathway

Figure 5 represents the interconnection between classes and instances in pentose phosphate pathway. In this figure we have shown classes and instances in the context of the Biopax ontology and the interconnectivity between instances defined by properties of the ontology.

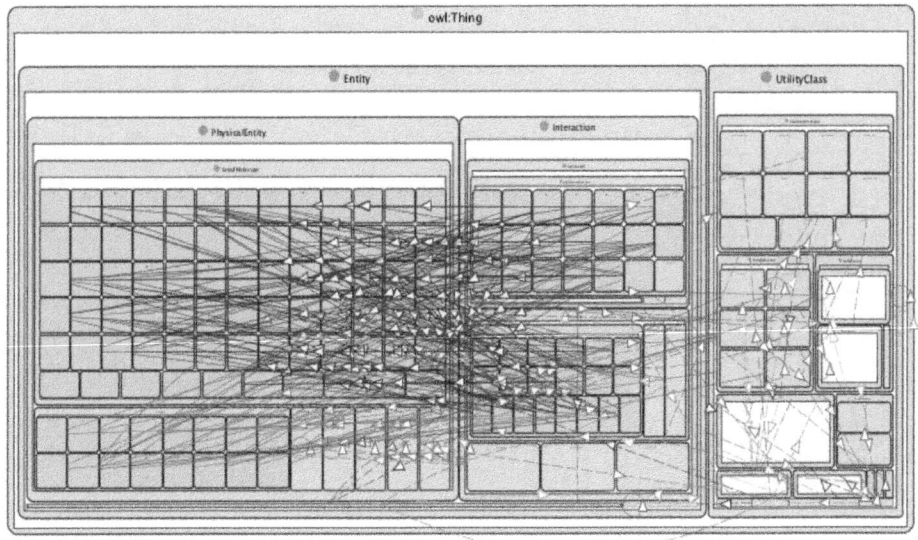

Fig. 5. Interconnectivity between classes and instances in pentose phosphate pathway in Biopax Level 3 ontology. Classes are represented in yellow labeled square, instances are represented in blue square and arrows represent interconnectivity between instances and classes.

On the other hand, each pathway contains a set of reactions that have been translated to Biochemical reactions subclass of Conversion class and Catalysis subclass of Control class (Figure 6). Participating compounds in biochemical reactions have been translated to Proteins subclass of Physical entity class and Small molecule subclass of Physical entity class. Finally, we have translated information using Biosource subclass, a subclass of Utility class and Xref subclass of Utility class in BiopaX Level 3.

We have used SB-KOM mediator to obtain mappings from KEGG base data information to Biopax Level 3 ontology. For example, steroid biosynthesis pathway of *Homo sapiens* (http://www.genome.jp/dbget-bin/www_bget?ko00100) is an instance of pathway subclass of entity class in Biopax Level 3 ontology and an instance of pathway class of AMMO ontology; metabolites involved in this pathway are instances of small molecules class, a subclass of physical entity and enzymes are instances of class protein, a subclass of physical entity in Biopax Level 3 ontology; chemical reactions of steroid biosynthesis are instances of biochemical reaction subclass and catalysis subclass in Biopax Level 3.

```xml
- <Pathway rdf:ID="UA141">
  - <pathwayComponent>
    - <Catalysis rdf:ID="UA61">
      - <controller>
        + <Protein rdf:ID="UA22"></Protein>
        </controller>
        <rdf:type rdf:resource="#BiochemicalReaction"/>
        <displayName rdf:datatype="http://www.w3.org/2001/XMLSchema#string">R03270</displayName>
      - <participant>
        - <SmallMolecule rdf:ID="UA62">
          <standardName rdf:datatype="http://www.w3.org/2001/XMLSchema#string">Thiamin diphosphate</standardName>
          <displayName rdf:datatype="http://www.w3.org/2001/XMLSchema#string">C00068</displayName>
          </SmallMolecule>
        </participant>
      - <participant>
        - <SmallMolecule rdf:ID="UA63">
          <displayName rdf:datatype="http://www.w3.org/2001/XMLSchema#string">C05125</displayName>
          <standardName rdf:datatype="http://www.w3.org/2001/XMLSchema#string">2-(alpha-Hydroxyethil)thiamine diphosphate</standardName>
          </SmallMolecule>
        </participant>
      - <participant>
        - <SmallMolecule rdf:ID="UA64">
          <standardName rdf:datatype="http://www.w3.org/2001/XMLSchema#string">Dihydrolipoyllysine-residue acetytransferase</standardName>
          <displayName rdf:datatype="http://www.w3.org/2001/XMLSchema#string">C16255</displayName>
          </SmallMolecule>
        </participant>
      - <participant>
        - <SmallMolecule rdf:ID="UA65">
          <standardName rdf:datatype="http://www.w3.org/2001/XMLSchema#string">Enzyme N6-(lipoyl)lysine</standardName>
          <displayName rdf:datatype="http://www.w3.org/2001/XMLSchema#string">C15972</displayName>
          </SmallMolecule>
        </participant>
      </Catalysis>
    </pathwayComponent>
  + <pathwayComponent></pathwayComponent>
  + <pathwayComponent></pathwayComponent>
  + <pathwayComponent></pathwayComponent>
  </Pathway>
```

Fig. 6. Representation of glycolysis/gluconeogenesis pathway divided in components that can contain several participants

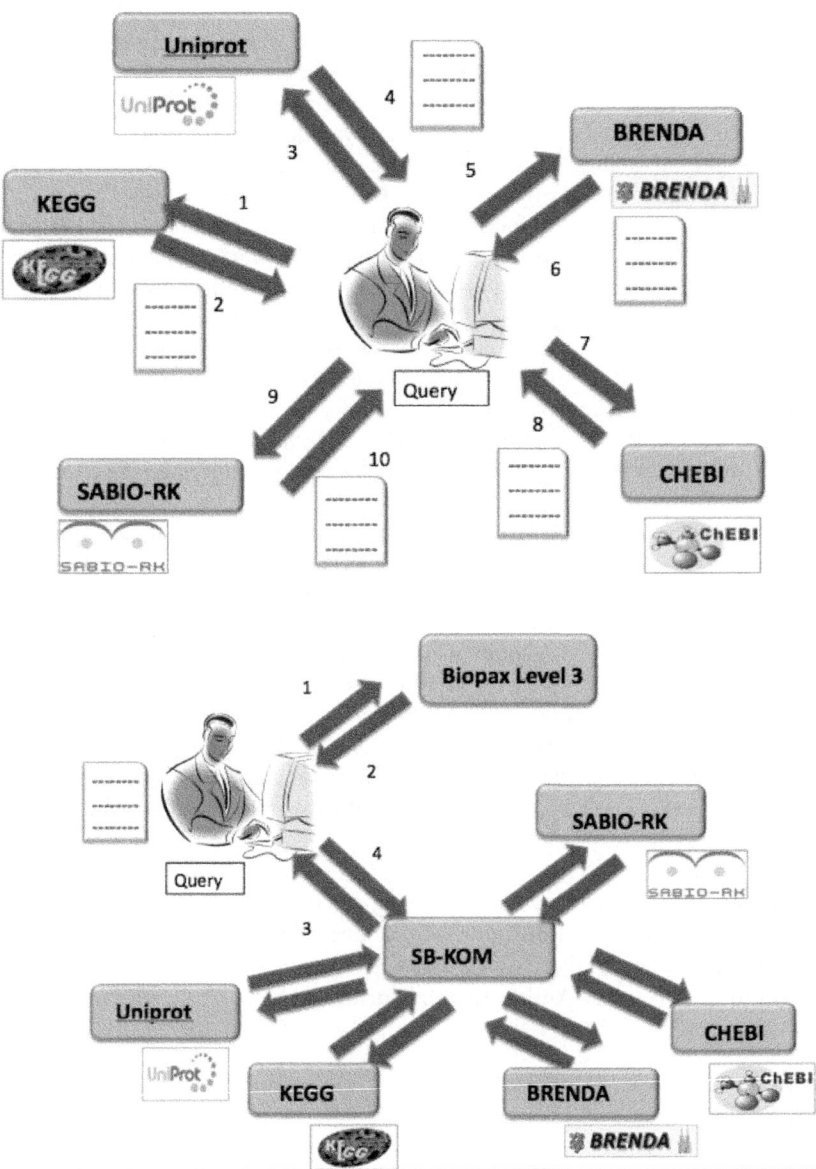

Fig.7. Superior image shows the process of mapping manually and inferior image shows the process of mapping by SB-KOM mediator. Numbers represetns the steps of each task.

5   Discussion and Conclusions

In this paper we have presented an analysis of similarities between two existing ontologies. The first ontology, AMMO, is an ad hoc ontology successfully used to integrate data of metabolic pathways. The second case, BioPax Level 3, is the best-known ontology to semantically represent pathway data. In this paper we present and approach to take advantage of the knowledge adquired in the design of AMMO, to enable the integration of pathway data using BioPax Level 3.

The use of previous knowledge reduces the cost of producing a mediator able to interpret queries in terms of BioPax and to produce BioPax instances. This work has been tested through four use cases corresponding to four well-known pathways.

These results obtained by SB-KOM mediator aims to optimize the process of creation of instances in Biopax level 3. If we wanted to solve this task manually, it is necessary to access each molecular biology database, query it by means of the user interface, interpret the data obtained, and finally combine these data to get some results. The use of the mediator, SB-KOM, will help biologists to extract large amounts of information from distributed databases and to automatically integrate and classify this information.

Figure 7 represents the differences of the creation of instances manually and automatically by using SB-KOM mediator. In Figure 7, the superior image shows the manual process to obtain information from some databases; a user must extract the information from data schemas and establish the correspondences of the concepts of Biopax Level 3 ontology to obtain information by using a biological query. The inferior shows one the simplicity of the process by using SB-KOM mediator; a user can realize a query and obtain the information in a semantic and automatic way.

Future work includes a complete test of the mediator produced using BioPax Level 3 as domain ontology, and the development of a user interface to allow users to take advantage of this system to access integrated pathway data and it could be graphically visualized.

Acknowledgments. This work has been supported by the Project Grants TIN2008-04844 (Spanish Ministry of Education and Science) and P07-TIC-02978 (Innovation, Science and Enterprise Ministry of the Junta de Andalucía.

We would like to thank Amine Kerzazi for his technical help on the development of the software.

References

1. Baxevanis, A.D., Ouellette, B.F.F.: Bioinformatics: a Practical Guide to the Analysis of Genes and Proteins. John Wiley and Sons, New York (2001)
2. Krieger, C.J., Zhang, P., Mueller, L.A., Wang, A., Paley, S., Arnaud, M., Pick, J., Rhee, S.Y., Karp, P.D.: Metacyc: a multiorganism database of metabolic pathways and enzymes. Nucleic Acids Research 32, 348 (2004)
3. Alfarano, C., Andrade, C.E., Anthony, K., Bahroos, N., Bayec, M., Bantoft, K., Betel, D., Bobechko, B., Boutilier, K., Burguess, E., et al.: The Biomolecular interaction network database and related tools 2005 update. Nucleic Acids Research 33, 418 (2005)

4. Demir, E., Babur, O., Dogrusoz, U., Gursoy, A., Nisanci, G., Cetin-Atalay, R., Orturk, M.: PATIKA: an Integrated Visual Enviroment for Collaborative construction and analysis of cellular pathaways. Bioinformatics 18, 996 (2002)
5. Joshi-Tope, G., Gillespie, M., Vastrik, I., D' Eustachio, P., Schmidt, E., De Bono, B., Jassal, B., Gopinath, G.R., Wu, G.R., Mathews, L., et al.: Reactome: a Knowledge Base of Biological Pathways. Nucleic Acid Research 33, 428 (2005)
6. Lemer, C., Antezana, E., Couche, F., Fays, F., Santolaria, X., Janky, R., Deville, Y., Richelle, J., Wodak, S.J.: The aMAZE, LightBench: a Web Interface to a Relational Database of Cellular Processes. Nucleic Acid Research 32, 443 (2004)
7. Kanehisa, M., Goto, S., Kawashima, S., Okuno, Y., Hattori, M.: The KEGG Resource for Deciphering the Genome. Nucleic Acid Research 32 (2004)
8. Alberts, B.: Molecular biology of the cell. Garland Science, New York (2002)
9. Schaefer, C.F., Anthony, K., Krupa, S., Buchoff, J., Day, M., Hannay, T., Buetow, K.H.: PID.: The Pathway Interaction Database. Nucleic Acids Research 37, 674–679 (2009)
10. Mi, H., Lazareva-Ulitsky, B., Loo, R., Kejariwal, A., Vandergriff, J., Rabkin, S., Guo, N., Muruganujan, A., Doremieux, O., Campbell, M.J., et al.: The PANTHER database of proteins families, subfamilies, functions and pathway. Nucleics Acid Research 33 (2005)
11. Navas Delgado, I., Montañez, R., Pino Ángeles, A., Moya García, A., Urdiales, J.L., Sánchez Jiménez, F., Aldana Montes, J.F.: AMMO-Prot: Amine System Project 3D-Model Finder. BMC Bioinformatics 9, S5 (2008)
12. Navas-Delgado, I., Aldana-Montes, J.F.: Extending SD-Core for Ontology-based Data Integration. J-Jucs 15, 3201–3230 (2009)
13. Reyes-Palomares, A., Montañez, R., Real-Chicharro, A., Chiniber, O., Kerzazi, A., Navas-Delgado, I., Medina, M.A., Aldana-Montes, J., Sánchez-Jiménez, F.: System Biology Metabolic Assitant: an ontology-based tool for integration of metabolic data in kinetic modeling. Bioinformatics 25, 834–835
14. Navas-Delgado, I., Montañez, R., Pino-Ángeles, A., Moya-García, A., Urdiales, J.L., Sánchez-Jiménez, F., Aldana-Montes, J.F.: AMMO-Prot: ASP Model Finder. BMC Bioinformatics 9, 1471–2105 (2008)
15. Chang, A., Scheer, M., Grote, A., Schomburg, I., Schomburg, D.: BRENDA, AMENDA and FRENDA the enzyme information system: new content and tools in 2009. Nucleic Acids Res. 37, 588–592 (2009)
16. Degtyarenko, K., De Matos, P., Ennis, M., Hastings, J., Zbinden, M., McNaught, A., Alcantara, R., Darsow, M., Guedj, M., Ashburner, M.: ChEBI: a database and ontology for chemical entities of biological interest. Nucleic Acids Res. 36, 344–350 (2008)
17. Bairoch, A., Apweiler, R., Wu, C., Barker, W., Boeckmann, B., Ferro, S., Gasteiger, E., Huang, H., Lopez, R., Magrane, M., et al.: The Universal Protein Resource (UniProt). Nucleic Acids Res. 33, 154–159 (2005)
18. Wittig, U., Golebiewski, M., Kania, R., Krebs, O., Mir, S., Weidemann, A., Anstein, S., Saric, J., Rojas, I.: SABIO-RK: Integration and Curation of Reaction Kinetics Data. In: Leser, U., Naumann, F., Eckman, B. (eds.) DILS 2006. LNCS (LNBI), vol. 4075, pp. 94–103. Springer, Heidelberg (2006)

Appendix

This table represents mappings between properties and classes of data schemes and Biopax Level 3 in SB-KOM mediator. No correspondences are represented in red.

Kegg Data Base

/Result/pathway/title	Pathway
/Result/pathway/title AND /Result/pathway/name	Pathway AND displayName
/Result/pathway/title AND /Result/pathway/title	Pathway AND standardName
/Result/pathway/title AND /Result/pathway/relatedPathways/path/name	Pathway AND relatedPathways
/Result/pathway/org	BioSource
/Result/pathway/org AND /Result/pathway/org	BioSource AND displayName
/Result/pathway/participantCompounds/cpd/name	SmallMolecule
/Result/pathway/participantCompounds/cpd/name AND /Result/pathway/participantCompounds/cpd/name	SmallMolecule AND displayName
/Result/pathway/participantCompounds/cpd/name AND /Result/pathway/participantCompounds/cpd/title	SmallMolecule AND standardName
/Result/pathway/reactions/reaction/name	BiochemicalReaction
/Result/pathway/reactions/reaction/name AND /Result/pathway/reactions/reaction/name	BiochemicalReaction AND displayName
/Result/pathway/reactions/reaction/name AND /Result/pathway/reactions/reaction/title	BiochemicalReaction AND standardName
/Result/pathway/reactions/reaction/name AND /Result/pathway/reactions/reaction/equation	BiochemicalReaction AND Stoichiometry
/Result/pathway/reactions/reaction/enzyme/ec	Protein
/Result/pathway/reactions/reaction/enzyme/ec AND /Result/pathway/reactions/reaction/enzyme/ec	Protein AND displayName
/Result/pathway/reactions/reaction/enzyme/ec AND /Result/pathway/reactions/reaction/enzyme/title	Protein AND standardName
/Result/pathway/reactions/reaction/enzyme/ec AND /Result/pathway/org	Protein AND BioSource AND organism
/Result/pathway/reactions/reaction/name AND /Result/pathway/reactions/reaction/enzyme/ec	BiochemicalReaction AND Protein AND controller
/Result/pathway/reactions/reaction/substrate/name	SmallMolecule
/Result/pathway/reactions/reaction/substrate/name AND /Result/pathway/reactions/reaction/substrate/name	SmallMolecule AND displayName
/Result/pathway/reactions/reaction/substrate/name AND /Result/pathway/reactions/reaction/substrate/title	SmallMolecule AND standardName
/Result/pathway/reactions/reaction/product/name	SmallMolecule
/Result/pathway/reactions/reaction/product/name AND /Result/pathway/reactions/reaction/product/name	SmallMolecule AND displayName
/Result/pathway/reactions/reaction/product/name AND /Result/pathway/reactions/reaction/product/title	SmallMolecule AND standardName
/Result/pathway/reactions/reaction/name AND /Result/pathway/reactions/reaction/substrate/name	BiochemicalReaction AND SmallMolecule AND (left OR right)
/Result/pathway/reactions/reaction/name AND /Result/pathway/reactions/reaction/product/name	BiochemicalReaction AND SmallMolecule AND (left OR right)
/Result/pathway/relatedPathways/path	Pathway
/Result/pathway/relatedPathways/path AND /Result/pathway/relatedPathways/path/name	Pathway AND displayName
/Result/pathway/relatedPathways/path AND /Result/pathway/relatedPathways/path/title	Pathway AND standardName

/Result/pathway/title AND /Result/pathway/org	Pathway AND BioSource AND organism
/Result/pathway/title AND /Result/pathway/participantCompounds/cpd/name	Pathway AND SmallMolecule AND (left OR right)
/Result/pathway/title AND /Result/pathway/reactions/reaction/name	Pathway AND BiochemicalReaction AND pathwayComponent
/Result/pathway/title AND /Result/pathway/relatedPathways/path	Pathway AND Pathway AND isConectedTo
/Result/reaction/name	BiochemicalReaction
/Result/reaction/name AND /Result/reaction/name	BiochemicalReaction AND displayName
/Result/reaction/name AND /Result/reaction/title	BiochemicalReaction AND standardName
/Result/reaction/name AND /Result/reaction/enzyme/ec	BiochemicalReaction AND controller
/Result/reaction/name AND /Result/reaction/path/name	BiochemicalReaction AND pathwayComponent
/Result/reaction/name AND /Result/reaction/equation	BiochemicalReaction AND Stoichiometry
/Result/reaction/substrate/name	SmallMolecule
/Result/reaction/substrate/name AND /Result/reaction/substrate/name	SmallMolecule AND displayName
/Result/reaction/substrate/name AND /Result/reaction/substrate/title	SmallMolecule AND standardName
/Result/reaction/product/name	SmallMolecule
/Result/reaction/product/name AND /Result/reaction/product/name	SmallMolecule AND displayName
/Result/reaction/product/name AND /Result/reaction/product/title	SmallMolecule AND standardName
/Result/reaction/name AND /Result/reaction/substrate/name	BiochemicalReaction AND SmallMolecule AND (left OR right)
/Result/reaction/name AND /Result/reaction/product/name	BiochemicalReaction AND SmallMolecule AND (left OR right)
/Result/reaction/path	Pathway
/Result/reaction/path AND /Result/reaction/path/name	Pathway AND displayName
/Result/reaction/path AND /Result/reaction/path/title	Pathway AND standardName
/Result/reaction/name AND /Result/reaction/path	BiochemicalReaction AND Pathway AND pathwayComponent
/Result/reaction/name AND /Result/reaction/enzyme/ec	BiochemicalReaction AND Protein AND controller
/Result/reaction/enzyme/ec	Protein
/Result/reaction/enzyme/ec AND /Result/reaction/enzyme/ec	Protein AND displayName
/Result/cpd/name	SmallMolecule
/Result/cpd/name AND /Result/cpd/name	SmallMolecule AND displayName
/Result/cpd/name AND /Result/cpd/title	SmallMolecule AND standardName
/Result/cpd/name AND /Result/cpd/path/name	SmallMolecule AND pathwayComponent
/Result/cpd/name AND /Result/cpd/enzyme/ec	SmallMolecule AND controller
/Result/cpd/reaction/name	BiochemicalReaction
/Result/cpd/reaction/name AND /Result/cpd/reaction/name	BiochemicalReaction AND displayName
/Result/cpd/name AND /Result/cpd/reaction/name	SmallMolecule AND BiochemicalReaction AND participant
/Result/cpd/path	Pathway

/Result/cpd/path AND /Result/cpd/path/name	Pathway AND displayName
/Result/cpd/path AND /Result/cpd/path/title	Pathway AND standardName
/Result/cpd/name AND /Result/cpd/path	SmallMolecule AND Pathway AND pathwayComponent

Chebi Data Base

/Result/main	SmallMolecule
/Result/main;/Result/main/ChEBI_Name	SmallMolecule;standardName
/Result/main;/Result/main/ChEBI_ID	SmallMolecule;displayName
/Result/main;/Result/main/SMILES	SmallMolecule;ChemicalStructure
/Result/main;/Result/main/MASS	SmallMolecule;SmallMoleculeReference; molecularWeight
/Result/main;/Result/main/CHARGE	SmallMolecule;charge
/Result/main;/Result/main/DataBase_Links/DataB_kegg/Database_Link	SmallMolecule;keggCompoundId
/Result/main;/Result/main/Registry_Numbers	SmallMolecule;casNumber
/Result/main;/Result/main/Formulas/Form	SmallMolecule;SmallMoleculeReference; formula
/Result/Automatic_Xrefs/Reactions_Pathways	BiochemicalReaction
/Result/Automatic_Xrefs/Reactions_Pathways;/Result/Automatic_Xrefs/Reactions_Pathways/SABIO_RK_S/SABIO_RK/Identifiers	BiochemicalReaction;displayName
/Result/Automatic_Xrefs/Reactions_Pathways;/Result/Automatic_Xrefs/Reactions_Pathways/SABIO_RK_S/SABIO_RK/Name	BiochemicalReaction;Stoichiometry
/Result/main;/Result/Automatic_Xrefs/Reactions_Pathways	SmallMolecule;BiochemicalReaction; participant
/Result/Automatic_Xrefs/Reactions_Pathways/SABIO_RK_S/SABIO_RK/Sustrate	SmallMolecule
/Result/Automatic_Xrefs/Reactions_Pathways/SABIO_RK_S/SABIO_RK/Sustrate;/Result/Automatic_Xrefs/Reactions_Pathways/SABIO_RK_S/SABIO_RK/Sustrate	SmallMolecule;standardName
/Result/Automatic_Xrefs/Reactions_Pathways/SABIO_RK_S/SABIO_RK/Product	SmallMolecule
/Result/Automatic_Xrefs/Reactions_Pathways/SABIO_RK_S/SABIO_RK/Product;/Result/Automatic_Xrefs/Reactions_Pathways/SABIO_RK_S/SABIO_RK/Product	SmallMolecule;standardName
/Result/Automatic_Xrefs/Reactions_Pathways;/Result/Automatic_Xrefs/Reactions_Pathways/SABIO_RK_S/SABIO_RK/Sustrate	BiochemicalReaction;SmallMolecule;(left OR right)
/Result/Automatic_Xrefs/Reactions_Pathways;/Result/Automatic_Xrefs/Reactions_Pathways/SABIO_RK_S/SABIO_RK/Product	BiochemicalReaction; SmallMolecule;(left OR right)

Uniprot Data Base

/Result/polypeptides/polypeptides_name	Protein
/Result/polypeptides/polypeptides_name AND /Result/polypeptides/uniprotId	Protein AND displayName
/Result/polypeptides/polypeptides_name AND /Result/polypeptides/synonym_polypeptides_name	Protein AND proteinSynonym
/Result/polypeptides/polypeptides_name AND /Result/polypeptides/polypeptides_name	Protein AND standardName
/Result/polypeptides/polypeptides_name AND /Result/polypeptides/primAc	Protein AND displayName

/Result/polypeptides/polypeptides_name /Result/polypeptides/ec_number	AND	Protein AND displayName
/Result/polypeptides/organism_name		BioSource
/Result/polypeptides/organism_name /Result/polypeptides/organism_name	AND	BioSource AND standardName
/Result/polypeptides/organism_name /Result/polypeptides/Taxon	AND	BioSource AND taxonXref
/Result/polypeptides/amino_acid_sequence		Protein OR SmallMolecule
/Result/polypeptides/amino_acid_sequence /Result/polypeptides/fasta_file	AND	(Protein OR SmallMolecule) AND fastaFile
/Result/polypeptides/polypeptides_name /Result/polypeptides/organism_name	AND	Protein AND BioSource AND organism
/Result/polypeptides/polypeptides_name /Result/polypeptides/amino_acid_sequence	AND	Protein AND (Protein OR SmallMolecule) AND hasA

GENote v.β: A Web Tool Prototype for Annotation of Unfinished Sequences in Non-model Eukaryotes

Noé Fernández-Pozo[1], Darío Guerrero-Fernández[2], Rocío Bautista[2],
Josefa Gómez-Maldonado[1], Concepción Avila[1], Francisco M. Cánovas[1],
and M. Gonzalo Claros[1,2]

[1] Departamento de Biología Molecular y Bioquímica, Universidad de Málaga,
Campus de Teatinos, 29071 Málaga, Spain
{noefp,pgomez,cavila,canovas,claros}@uma.es
http://www.bmbq.uma.es/fmp
[2] Plataforma Andaluza de Bioinformática, Universidad de Málaga,
Severo Ochoa 34, 29590 Málaga, Spain
{dariogf,rociobm,claros}@scbi.uma.es
http://www.scbi.uma.es/pab

Abstract. *De novo* identification of genes in newly-sequenced eukaryotic genomes is based on sensors, which are not available in non-model organisms. Many annotation tools have been developed and most of them require sequence training, computer skills and accessibility to sufficient computational power. The main need of non-model organisms is finding genes, transposable elements, repetitions, etc., in reliable assemblies. GENote v.β is intended to cope with these aspects as a web tool for researchers without bioinformatics skills. It facilitates the annotation of new, unfinished sequences with descriptions, GO terms, EC numbers and KEEG pathways. It currently localises genes and transposons, which enable the sorting of contigs or scaffolds from a BAC clone, and reveals some putative assembly inconsistencies. Results are provided in GFF3 format and in tab-delimited text readable in viewers; a summary of findings is provided also as a PNG file.

Keywords: Annotation, web tool, unfinished sequence, gene finding, non-model species.

1 Introduction

Annotation is the process of interpreting raw genomic sequences into useful biological information by integrating computational analyses, auxiliary biological data and biological expertise. It should begin as early in a project as is possible, because the analysis of the assembled sequence will often identify problems in the raw sequence or in its assembly [2]. Genome annotation is best carried out by combining several methods, being very successful *cis*- and *trans*-alignments, and also the *de novo* gene prediction when a related species is well known (*de novo* gene predictors have repeatedly proven to be more challenging than expected) [2].

A.T. Freitas and A. Navarro (Eds.): JBI 2010, LNBI 6620, pp. 66–71, 2012.

In recent years, the biological community has started to see the dramatic impact of new sequencing technologies on the number of sequenced genomes, and it is expected that this influx of data will continue to escalate in the near future. Annotation of newly-sequenced eukaryotic genomes is based on signal sensors such as promoters, splice junctions, start and stop codons, branch points, polyadenylation sites or untranslated regions, as well as content sensors. Sensors can be predicted only for well known species and microorganisms [7], but use to produce incorrect predictions in newly-sequenced organisms [9], with more dramatic effects on plant genomes [2]. Genome annotation is therefore becoming the bottleneck in functional genomics today. Eukaryotic genomes are particularly at risk as their large size and intron-containing genes make them difficult targets for a straightforward annotation.

Many annotation tools have been developed and most of them require a cumbersome installation including installation of external executables, sequence training, programming skills and accessibility to sufficient computational power. Although many laboratories where organisms are sequenced lack these bioinformatics experience, they could annotate their datasets with specialised databases (like Ensembl or VectorBase) provided that they work with 'popular' (model) species. On the other hand, reliable annotated sequences had required a manual curation performed by trained "bio-curators" [8], and is only available for a few model species [7]. GO, KEGG and EC terms transfer based on sequence similarity is the most extended approach as it is probably the methodology that best suits the desired high-throughput and high-coverage needs of sequence annotation at a genomic scale, being Blast2GO [5] and AutoFact [10] among the most widely used software for that. Gene localisation can be performed with MAKER [4], FGENESH [12] or mGene.web [13], which require previous training with closely-related species. But, when trying to annotate eukaryotic non-model organisms, a number of drawbacks are encountered mainly due to a lack of significant homology between query sequences and database sequences [11]. Moreover, it is recognised that good performance of an annotator on an established genome is not guarantee of similar performance on an emerging genome [4]. We hope that GENote v.β can close the gap of annotation of newly-sequenced, non-model (and even unfinished) species using both DNA and protein databases as references, and can give clues about the contig arrangement and putative assembly errors. Examples of GENote v.β performance are given.

2   Implementation

GENote v.β (provisional address: http://www.scbi.uma.es/genotebeta) was built as a command line application in Ruby (http://www.ruby-lang.org/) and the web interface was given with InGeBiol (Guerrero-Fernández and Claros, in preparation), a Ruby-on-Rails framework that uses Ajax to facilitate the development of a web interface for any command-line software or pipeline. Internal configuration files are in JSON (http://www.json.org), a data format that competes with XML for highly human-readable syntax, and for efficiency in the storage and parsing phases. Similarity searches were customised using Blast+ [3]. Firefox or Safari web browsers are recommended. GENote v.β executions are queued to a 80 core cluster with 160 Gb of RAM.

GENote v.β architecture is modular, which allows addition of new annotation modules and easy parallelisation of the whole process. The input is only a genomic sequence in FASTA format since default values are given for the three required parameters. Results contain a summary GFF3 (also shown as a PNG image using the software AnnotationSketch [14]) and a series of detailed files also in GFF3 that can be directly viewed in genome browsers such as Apollo or AnnotationSketch. Since the current version is an unfinished prototype, interface and results are subjected to changes during the development of this tool.

3   Results and Discussion

Annotators can predict today a plethora of features in a genomic sequence in model species. However, when studying non-model organisms, the main goal is to discern the presence of a gene, the putative function of such gene, its promoter location, the putative intron-exon pattern or the presence of transposable elements. The orientation of such a gene and the presence of other genes close to it is also welcome. As examples of these needs, a standard approach has been recently applied to a small set of BAC sequences from pine (a gymnosperm) using MAKER trained with both dicot and monocot parameters [11] with quite unsatisfactory results; and two BAC clones from spruce (another gymnosperm) genes were analysed with FGENESH using the dicot matrix and manual curation [6]. In contrast, Fig. 1 shows the successful intron-exon pattern of unfinished BAC clones from our laboratory using GENote v.β. The result seems to be consistent using the contigs forming the BAC (Fig. 1A), as well as their scaffold (Fig. 1B). Researchers working on non-model organisms for which only partial assemblies or scaffolds are available will therefore find useful GENote v.β for analysing their ongoing sequencing, since it is well known that accuracy of annotators when evaluated on isolated genes is much better than their accuracy when evaluated on a whole genome [1].

Since the gold standard of annotation will continue to be a full-length cDNA sequence with a computationally inferred ORF [2], GENote v.β does not follow the sensor approach but makes use of hierarchical similarity searches against UniProt proteins and the non-redundant EMBL nucleotides to localise protein-coding genes, putative introns and exons, transposons and any other desired type of sequence (Fig. 1). It extracts descriptions, GO terms, EC numbers and KEEG pathways from the most similar (desirably orthologous) sequence. The way to ascertain if the alignment covers the entire transcription unit is based on the protein sequence when available; the rationale is that the number of residues in closely related members of a globular protein family usually falls within a relatively narrow range [7]. If sequences from ultra-high-throughput sequencing are available, they could be added to GENote in order to detect low-transcribed genes o species-specific genes.

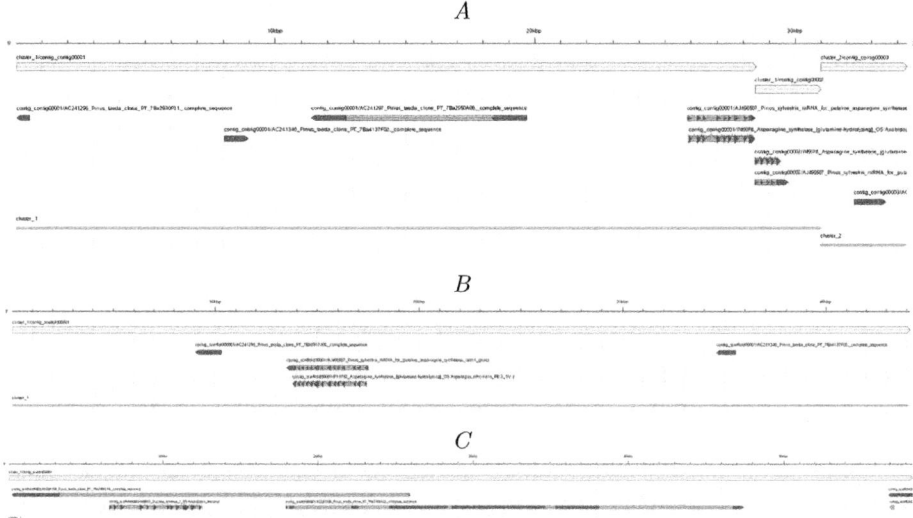

Fig. 1. Example of a global overview of unfinished BAC clones provided by GENote v.β. Box sizes indicate the sense of transcription. Red boxes are deduced from a protein homology using UniProt; blue boxes are deduced from DNA sequences using non-redundant EMBL database. Lower grey lines indicate the putative "super-contig" that can be predicted with the studied contigs. *A*, analysis of the three contigs generated from a single BAC clone of *Pinus pinaster* genomic DNA. The first two contigs were sorted in order to reconstruct the intron-exon pattern of a putative gene. *B*, analysis of a single scaffold covering the same BAC clone than in *A* has revealed the presence of the complete gene containing the whole exon set for that gene. The blue boxes on the right and on the left are supposed to be ESTs, but it is clear that these ESTs were obtained from genomic DNA and not from cDNA. *C*, analysis of the scaffold of another BAC clone revealing that there is a putative gene deduced only from a similar protein sequence; there are also long stretches of interrupted DNAs that could be due to heterogeneity among species or a misassembly, both reflecting that the same EMBL subject sequence (blue boxes contained in the same grey box) contain similarities to discontinuous BAC sequences.

Although GENote v.β could be used for automated annotation of any nucleotide sequence, it is mainly focused on genomic DNAs, even if they have gaps of unknown size (scaffolds) that can be considered unfinished or draft-quality assemblies from a single BAC clone. In such a case, GENote v.β is able of sorting contigs to form a "super-contig" provided that they share discontinuous fragments of the same putative subject sequence. One can make use of this capability to localise reliable and suspicious assemblies among the contigs/scaffolds analysed (Fig. 1C): the clue are genomic sequences in nucleotide database that are interrupted by non-homologous sequences, which could indicate that the sequence corresponding to such DNA is repeated or misassembled.

Annotation results are provided in GFF3 format and in tab-delimited text with the order and contig orientations into the "super-contigs". These files can be imported and managed in a database or spreadsheet and combined with existing annotations as desired. Graphic viewers, such as AnnotationSketch, Apollo or Artemis, can be used for displaying the annotation results, although only the first one seems to read without problems the complete specifications of GFF3 formatted files.

4   Conclusion

GENote v.β is a generic tool prototype based on sequence similarity that can be used for automated annotation of any nucleotide sequences (prokaryotic and eukaryotic), although it is mainly focused on genomic DNAs even if they have gaps of unknown size, that is, contig or scaffolds of unfinished, or draft-quality, assemblies. It is able to sort contigs belonging to the same gene or suggest a misassembly problem. GENote v.β does not need configuration nor installation nor training nor bioinformatics skills to be run.

In order to obtain a finished bioinformatics tool, some additional features are ongoing, such as (i) accurate intron-exon boundaries, (ii) localisation of complete transposable elements, (iii) repetitive sequence pattern, (iv) presence of unwanted sequences provided by the user, and (v) parallelisation of the Ruby script (since the BLAST+ calls are already executed in parallel).

Acknowledgments. The authors gratefully acknowledge the computer resources and technical support provided by the Plataforma Andaluza de Bioinformática of the University of Málaga, Spain. This study was supported by grants from the Spanish Ministerio de Ciencia e Innovación [BIO2009-07490], the European Union [PLE2009-0016], and the Junta de Andalucía [CVI-6075].

References

1. Brent, M.R.: Genome annotation past, present, and future: how to define an orf at each locus. Genome Res. 15(12), 1777–1786 (2005)
2. Brent, M.R.: Steady progress and recent breakthroughs in the accuracy of automated genome annotation. Nat. Rev. Genet. 9(1), 62–73 (2008)
3. Camacho, C., Coulouris, G., Avagyan, V., Ma, N., Papadopoulos, J., Bealer, K., Madden, T.L.: Blast+: architecture and applications. BMC Bioinformatics 10, 421 (2009)
4. Cantarel, B.L., Korf, I., Robb, S.M.C., Parra, G., Ross, E., Moore, B., Holt, C., Sánchez Alvarado, A., Yandell, M.: Maker: an easy-to-use annotation pipeline designed for emerging model organism genomes. Genome Res. 18(1), 188–196 (2008)
5. Götz, S., García-Gómez, J.M., Terol, J., Williams, T.D., Nagaraj, S.H., Nueda, M.J., Robles, M., Talón, M., Dopazo, J., Conesa, A.: High-throughput functional annotation and data mining with the blast2go suite. Nucleic Acids Res. 36(10), 3420–3435 (2008)

6. Hamberger, B., Hall, D., Yuen, M., Oddy, C., Hamberger, B., Keeling, C.I., Ritland, C., Ritland, K., Bohlmann, J.: Targeted isolation, sequence assembly and characterization of two white spruce (picea glauca) bac clones for terpenoid synthase and cytochrome p450 genes involved in conifer defence reveal insights into a conifer genome. BMC Plant Biol. 9, 106 (2009)
7. Harrow, J., Nagy, A., Reymond, A., Alioto, T., Patthy, L., Antonarakis, S.E., Guigó, R.: Identifying protein-coding genes in genomic sequences. Genome Biol. 10(1), 201 (2009)
8. Jones, C.E., Brown, A.L., Baumann, U.: Estimating the annotation error rate of curated go database sequence annotations. BMC Bioinformatics 8, 170 (2007)
9. Korf, I.: Gene finding in novel genomes. BMC Bioinformatics 5, 59 (2004)
10. Koski, L.B., Gray, M.W., Lang, B.F., Burger, G.: Autofact: an automatic functional annotation and classification tool. BMC Bioinformatics 6, 151 (2005)
11. Kovach, A., Wegrzyn, J.L., Parra, G., Holt, C., Bruening, G.E., Loopstra, C.A., Hartigan, J., Yandell, M., Langley, C.H., Korf, I., Neale, D.B.: The pinus taeda genome is characterized by diverse and highly diverged repetitive sequences. BMC Genomics 11, 420 (2010)
12. Salamov, A.A., Solovyev, V.V.: Ab initio gene finding in drosophila genomic dna. Genome Res. 10(4), 516–522 (2000)
13. Schweikert, G., Behr, J., Zien, A., Zeller, G., Ong, C.S., Sonnenburg, S., Rätsch, G.: mgene.web: a web service for accurate computational gene finding. Nucleic Acids Res. 37(Web Server issue), W312–W316 (2009)
14. Steinbiss, S., Gremme, G., Schärfer, C., Mader, M., Kurtz, S.: Annotationsketch: a genome annotation drawing library. Bioinformatics 25(4), 533–534 (2009)

Interface Similarity Improves Comparison of DNA-Binding Proteins: The Homeobox Example

Álvaro Sebastián[1], Carlos P. Cantalapiedra[1], and Bruno Contreras-Moreira[1,2]

[1] Laboratorio de Biología Computacional,
Estación Experimental de Aula Dei/CSIC,
Av. Montañana 1005, Zaragoza, España
[2] Fundación ARAID, Paseo María Agustín 36, Zaragoza, España
http://www.eead.csic.es/compbio
{asebastian,bcontreras}@eead.csic.es

Abstract. The recently published 3D-footprint database contains an up-to-date repository of protein-DNA complexes of known structure that belong to different superfamilies and bind to DNA with distinct specificities. This repository can be scanned by means of sequence alignments in order to look for similar DNA-binding proteins, which might in turn recognize similar DNA motifs. Here we take the complete set of Homeobox proteins from *Drosophila melanogaster* and their preferred DNA motifs, which would fall in the largest 3D-footprint superfamily and were recently characterized by Noyes and collaborators, and annotate their interface residues. We then analyze the observed amino acid substitutions at equivalent interface positions and their effect on recognition. Finally we estimate to what extent interface similarity, computed over the set of residues which mediate DNA recognition, outperforms BLAST expectation values when deciding whether two aligned Homeobox proteins might bind to the same DNA motif.

Keywords: protein-DNA interface, DNA motif, substitution matrices.

1 Introduction

3D-footprint [1] (http://floresta.eead.csic.es/3dfootprint) is a database that dissects sequence readout in protein-DNA complexes of known structure, extracted from the Protein Data Bank [2], identifying molecular contacts that contribute to specific recognition and inferring structure-based position weight matrices from the atomic coordinates. Currently the database contains over 2700 complexes, which can be assigned to SCOP superfamilies [3]. After removing redundancy, the most populated superfamily turns out to be that of homeodomain-like proteins, including Homeobox transcription factors, which have been the subject of extensive crystallographic and spectroscopic studies due to their key role in developmental processes in multicellular organisms [4].

A.T. Freitas and A. Navarro (Eds.): JBI 2010, LNBI 6620, pp. 72–82, 2012.

Furthermore, Homeobox proteins are of special interest since the publication of the work by Noyes and collaborators [5], in which the authors characterized the binding specificities of 85 *Drosophila melanogaster* homeodomains. This repertoire of homologous transcription factors provides a formidable opportunity to study the correlation between the mutations that naturally occur at the interface of Homeobox proteins and their effect on binding specificity.

In this paper we apply the structural knowledge contained in 3D-footprint to: i) define the set of most commonly used interface residues across Homeobox proteins; and ii) elucidate to what extent interface similarity between pairs of homeodomains correlates with the recognition of similar DNA motifs.

After a cross-validation benchmark we find that interface position-specific substitution matrices (ISUMs), automatically inferred from training sets of homeodomains, perform better than BLOSUM62, and significantly better than BLAST expectation values, in the task of deciding whether two aligned Homeobox proteins bind to the same DNA motif.

2   Material and Methods

2.1   Homeobox Protein Sequences and DNA Motifs

A dataset of 85 *D.melanogaster* Homeobox protein sequences and their 2240 DNA binding sites, first published by Noyes [5], was used to build 85 position weight matrices (PWMs) in TRANSFAC format using both CONSENSUS [6] and MEME [7], choosing in each case the resulting PWM with largest information content.

2.2   Structural Alignment of Homeodomains and Identification of Interface Residues

A multiple structural alignment of non-redundant homeodomains extracted from 3D-footprint [1] was compiled as previously explained [10], and all identified interface interactions annotated as hydrogen bonds, water-mediated hydrogen bonds or hydrophobic interactions. For simplicity interface residues were numbered following the schema using by Noyes [5]. The list of annotated homeodomains includes 35 Protein Data Bank chains:

```
1au7_B_1,1b72_A_1,1b8i_A_1,1b8i_B_1,1e3o_C_1,1fjl_C_1,1h89_C_1,1h89_C_2,
1hlv_A_2,1ic8_A_1,1ig7_A_1,1ign_A_1,1ign_A_2,1jgg_B_1,1jt0_C_1,1le8_A_1,
1mnm_C_1,1nk3_P_1,1puf_A_1,1puf_B_1,1w0t_B_1,1w0u_B_1,1yz8_P_1,1zq3_P_1,
2d5v_B_1,2h1k_B_1,2hdd_A_1,2kdz_A_1,2kdz_A_2,2qhb_A_1,2r5y_A_1,2yvh_C_1,
3cmy_A_1,3d1n_I_1,9ant_B_1
```

Furthermore, these homeodomains were sampled to calculate pairwise interface alignments. This procedure starts by reducing their protein-DNA interfaces to two-dimensional matrices, which we call *interface matrices*, that are expected to capture most details of their binding mode. Then, a pair of such matrices can be aligned by i) matching interface amino acid residues whose contact patterns overlap, and ii) by penalizing pairs of residues with distinct contact maps. As a by-product of these interface alignments we also obtain structure-based alignments of their bound DNA sequences, as shown in Figure 1.

2.3   Annotation of Interface Residues in Homeobox Proteins

The interface positions of all 85 *D.melanogaster* Homeobox protein sequences were assigned by means of local BLASTP [8] alignments to 3D-footprint entries.

2.4   DNA Motif Alignment and Similarity Scoring

All 85 *D.melanogaster* position weight matrices (PWMs, see Supplementary Material), which were generated with the DNA binding sites described in section 2.1, were aligned against each other with the STAMP software[9], using an ungapped Smith-Waterman algorithm and taking the Pearson Correlation Coefficient as the similarity score. This *similarity score* takes values in the range [-L , L], where L is the length of the PWM.

2.5   Cross-Validation Parameters

The original dataset was split into training and validation subsets of 68 and 17 homeodomains, respectively. This process was repeated for 10 rounds with different random training and validation sets. Training sets were used to compute ISUMs for each interface position, while validation sets were used to benchmark the DNA motif predictions made by applying the previously calculated ISUMs.

2.6   Generating Interface Substitution Matrices (ISUMs)

Homeobox domain sequences were globally aligned with MUSCLE [11] and their interface positions labelled. For each of the 8 interface positions, in bold in Table 1, the 4 most abundant amino acids were selected. All the 2^{10} possible binary score variations with repetitions among pairs of these 4 residues were computed. Obviously non binary scores are possible and probably more realistic, but at the cost of increasing the search space. For example, the four residues most frequently found in interface position 2 (G,R,Q,K) could be assigned the following 10 substitution scores, which represent the chance of mutating one residue to another while preserving the ability to recognize the same DNA motif: $GG \rightarrow 1, RR \rightarrow 1, GQ \rightarrow 0, KK \rightarrow 1, QR \rightarrow 0, KR \rightarrow 1, QQ \rightarrow 1, GK \rightarrow 0, KQ \rightarrow 0, GR \rightarrow 0$.

For each interface position the best score variations, those that maximized the Pearson correlation between interface scores and the corresponding DNA motif alignment scores, were selected and used to build symmetric interface substitution matrices (ISUMs). Ten sets of ISUMs were generated (one per training set) and used independently to perform DNA motif predictions within each of the corresponding validation sets. The final ISUMs in Tables 2 and 3 are the average of 10 cross-validations rounds.

2.7   Pairwise Alignments of Homeobox Domains

Pairs of protein sequences from the validation set where aligned with the BLASTP program [8]. Resulting E-values where normalized by taking the natural logarithm and dividing with respect to self-match E-values, as shown for sequences A and B in Equation1:

$$
normEval(A, B) = \begin{cases} 0 & if\ Eval(A, B) > 0; \\[2mm] 1 & if\ Eval(A, B) < AbsMinEv; \\[2mm] \frac{ln(Eval(A,B))}{ln(minEval))} & if\ minEval \geq AbsMinEv; \\[2mm] \frac{ln(Eval(A,B))}{ln(AbsMinEv)} & if\ minEval < AbsMinEv. \end{cases} \tag{1}
$$

where $minEval = min(Eval(A, A), Eval(B, B))$ and $AbsMinEv$ is the absolute smallest non-zero E-value returned by BLAST ($1e^{-180}$). Normalized E-values take values in the range 0-1, with small values corresponding to low sequence similarity.

2.8   Validation of the Calculated ISUMs

All possible pairwise alignments from the validation set, extracted from the global multiple alignment mentioned in section 2.6, were sampled and their interfaces scored with the trained ISUMs and with the generic substitution matrix BLOSUM62 [12]. In addition, all sequence pairs from the validation set were also re-aligned with BLASTP in order to calculate normalized E-values, which we used as a measure of overall sequence similarity. Data pairs of interface and motif alignment scores were scatter-plotted, linear regression estimated by least-squares fitting and statistical parameters calculated with statistical software R [13]. A receiver operating characteristic (ROC) curve was also plotted taking as *truth test* a motif *similarity score* ≥ 5 . This threshold was tuned after benchmarking the content of TRANSFAC database v9.3 [14], in order to obtain a sensitivity (True Positive Rate) of 0.7 and a specificity (1-False Positive Rate) of 0.9.

3   Results

3.1   Comparison and Clustering of Homeobox Interface Architectures

A non-redundant set of homeodomains included in 3D-footprint [1] was analyzed and their protein-DNA interfaces reduced to two-dimensional matrices. Subsequently these 2D interface matrices were compared to each other, end the corresponding DNA motif alignments extracted, as illustrated in Figure 1.

As a result of this structural analysis we found that Homeobox DNA motifs usually fit one of 7 subtypes, shown in Figure 2, which approximately encompass the 11 groups originally proposed by Noyes [5]. These clusters show the structural equivalence between different subtypes of Homeobox DNA motifs, which we consider in section 3.6 in order to call incorrect DNA alignments.

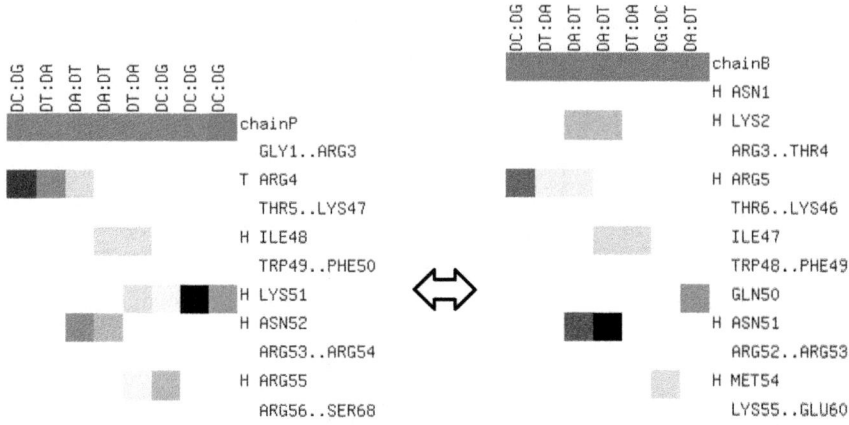

Fig. 1. Structure-based interface alignment of homeodomains 1zq3_P (left) and 2h1k_B (right). Each complex features interface residues in the Y axis and DNA nucleotides in the X axis (grey bar on top). Interactions are depicted as filled squares with density proportional to the number of atomic contacts. Aligned (equivalent) interface residues are placed in the same row. The resulting DNA motif alignment is: CTAATCCC / CTAATGA- .

3.2   Defining a Consensus Homeobox Protein-DNA Interface

Homeobox proteins (and in general homeodomain-like proteins) usually bind to DNA with a conserved architecture. This observation can be used to infer interface conservation directly from sequence alignments between transcription factors. However, a consensus set of residues which i) faithfully represent the interface architecture, and ii) minimize the loss of information, must be defined beforehand.

In the case of Homeobox proteins we used structural data collected in the 3D-footprint database [1] to annotate the critical interface residues involved in DNA recognition, which are shown in Table 1. Out of the 18 surveyed interface positions, which vary in terms of number of contacts and in frequency across homeodomain-like sequences, we shortlisted the same 8 positions proposed by Noyes [5] in order to facilitate the comparison of our results. These positions include 47, 50 and 54, which have been previously reported to be the key determinant positions for DNA recognition in mouse Homeobox transcription factors [15].

DNA motif

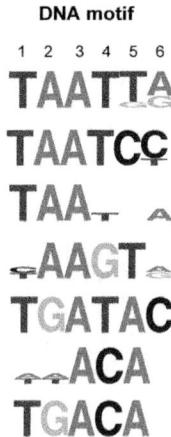

Fig. 2. Multiple alignment of Homeobox DNA motif subtypes, inferred from pairwise structure-based interface comparisons

Table 1. Survey of interface residues in homeodomains, numbered as in the work of Noyes [5]. The 8 most relevant residues for Homeobox proteins are in bold.

residue number	**2**	**3**	**4**	**5**	**6**	29	41	42	43	44	46	**47**	**50**	**51**	53	**54**	**55**	58
Hbonds	3	3	2	21	1	1	1		2		7	1	8	17	9	21	1	5	5	1
water-Hbonds	4								1			4	1	6	1				1
hydrophobic									1				8	2	1			

3.3 Derivation of Interface Substitution Matrices (ISUMs)

A 10-round cross-validation experiment was set up in order to calculate Interface Substitution Matrices (ISUMs) from randomly sampled training subsets of 68 Homeobox sequences. The resulting average matrices are shown in Table 2. ISUMs showed to be very similar from one training subset to another. In particular, inspection of the total 690 amino acid substitution scores generated (out of a theoretical maximum of 800: 10 per interface position x 8 positions x 10 rounds), only 160 scores differed from one training set to another and 80 belonged to positions 5 and 51. These later positions indeed contribute little to increase the correlation between the interface and the DNA recognized, as already envisaged by Noyes [5], and this might explain their variability.

As a control experiment, we also generated ISUMs for two randomly chosen non-interface positions (36 and 66, Table 3), which display an even higher variability in its amino acid composition and a lower correlation with the DNA motifs recognized. In general, any other sequence positions without interface roles contribute very little, if anything, to the DNA motif correlation. As a consequence, it is possible to increase the list of interface residues if necessary, as only relevant interface residues will have a correlation impact. In fact we observed this behaviour when using an enlarged list of 14 interface positions (data not shown).

Table 2. Average ISUMs for the 8 most important residues (in bold) of the Homeobox binding interface. The values in the matrix are average scores that evaluate the effect of mutating interface residues, measured in terms of DNA motif similarity score. Substitution scores take values in the range [0, 1]. A score close to 0 means that a substitution does not contribute to increase the correlation between interface similarity and DNA motif similarity across pairs of proteins in the training set.

2	A	E	K	R	**3**	A	H	K	R	**5**	Q	R	S	T	**47**	I	N	T	V
A	0.9	0	0	0	A	1	0	0	0	Q	0.5	0.9	0.15	0.2	I	1	0	1	1
E	0	0.8	0	0	H	0	1	0.9	1	R	0.9	1	0.05	0.1	N	0	0.1	0	0
K	0	0	1	1	K	0	0.9	0	0	S	0.15	0.05	0.5	0.85	T	1	0	1	1
R	0	0	1	1	R	0	1	0	1	T	0.2	0.1	0.85	0.5	V	1	0	1	1
50	A	I	K	Q	**51**	L	N			**54**	A	M	R	T	**55**	K	Q	R	
A	1	0	0	0	L	0.5	0.1			A	1	1	0	1	K	1	0.1	0	
I	0	0.85	0	0	N	0.1	0.9			M	1	1	0	1	Q	0.1	0.5	0.1	
K	0	0	0.2	0						R	0	0	0	0	R	0	0.1	0	
Q	0	0	0	1						T	1	1	0	1					

Table 3. Average ISUMs for two random non-interface positions (in bold) of the Homeobox binding interface. Note that these matrices contain more than 4 residues, as different cross-validation rounds often find a different set of frequent residues for these positions. Substitution scores take values in the range [0, 1].

36	A	H	K	M	N	Q	S	**66**	A	G	K	P	Q	S
A	0.1	0.4	0	0.1	0	0	0.1	A	0.7	0	0	0.6	0.2	0.3
H	0.4	0.7	0	0.1	0.1	0.1	0.1	G	0	0	0	0	0	0
K	0	0	0	0	0	0	0	K	0	0	0	0	0	0
M	0.1	0.1	0	0.2	0	0	0	P	0.6	0	0	0.1	0	0.3
N	0	0.1	0	0	0	0	0	Q	0.2	0	0	0	0	0
Q	0	0.1	0	0	0	0	0	S	0.3	0	0	0.3	0	0.3
S	0.1	0.1	0	0	0	0	0.3							

3.4   Evaluation of ISUMs

In each cross-validation round the derived ISUMs were evaluated on the collection of pairwise alignments of the remaining 17 protein sequences (the evaluation subset), annotating the 8 critical interface residues defined earlier. The correlations between interface similarity scores and the corresponding DNA motif similarities were calculated using the generated ISUMs and compared to those obtained using BLOSUM62 (for the comparison of interface residues) and the normalized E-value (for the complete protein sequences). Results are shown in Table 4.

In 4 of the 10 repetitions, the correlation obtained using ISUMs matrices showed to be at least 10% better than any other scoring, and in all but one repetitions it was at least 10% better than the normalized BLAST E-value. In addition, we measured the predictive power of interface comparison by means of a Receiver Operating Characteristic (ROC), plotted in Figure 3. The ROC

Table 4. Pearson correlation coefficients between Homeobox DNA motif and interface similarity using different scoring schemes in ten cross-validation rounds

round	ISUMs	BLOSUM62	normalized E-value
1	0.83	0.78	0.71
2	0.73*	0.59	0.45
3	0.71*	0.58	0.55
4	0.78	0.72	0.65
5	0.45	0.45	0.40
6	0.77	0.76	0.57
7	0.80*	0.71	0.64
8	0.66	0.74	0.42
9	0.82*	0.69	0.49
10	0.86	0.79	0.74

curve shows a significant improvement in the sensitivity and specificity when using ISUMs matrices in the range of False Positive Rate [0, 0.4] in comparison with the other measures, in particular when compared with BLAST expectation values.

3.5 Limitations of BLASTP Alignments When Predicting Homeobox DNA Motifs

Table 5 highlights 10 homeodomain pairs which display high DNA motif and interface similarities but low overall protein sequence similarity, and hence small normalized E-values. These alignments illustrate that often protein domain alignments, such as local alignments produced by BLASTP, might fail to explain the binding of similar DNA motifs. What is the frequency of these events? Among all the unique validation alignments (1138) there were 155 (13%) where normalized E-value was less than 0.2 and DNA motif similarity more than 0.6 . Out of these 155, 124 had an ISUM score higher than 0.6 and therefore demonstrate that there are a substantial amount of cases where DNA motifs can only be properly predicted taking into account interface similarity. Moreover, all of these 155 alignments have overall ISUMs scores higher than 0.2, so ISUMs clearly have a lower false negative rate than E-values.

Table 5. Example of Homeobox pairwise alignments with low overall protein similarity and high DNA motif and interface similarities. Motif similarity, ISUM and BLOSUM62 scores are normalized by dividing by motif length.

pair	DNA Motifs	Motif similarity	Interface	norm.E-value	ISUM	BLOSUM62
Lim1 / Slou	kyTaATTr/yaATTAam	0.91	RGRVQNSK/RRRIQNTK	0.1	0.75	0.71
Ro / CG4136	gyTAATTA/yAATTArs	0.85	RRRIQNAK/RHRVQNAK	0	1	0.78
Bsh / CG32105	gyymATTA/yTAATTAaw	0.83	RKRTQNMK/KRRVQNAK	0.1	0.88	0.51
H2.0 / Lim3	vkwTwATwAA/vyTAATTA	0.77	SWRVQNMK/KRRVQNAK	0.05	0.75	0.56
Ap / CG15696	TmATTArs/btTAATTr	0.75	KRRVQNAK/RLRIQNAR	0.06	0.75	0.65

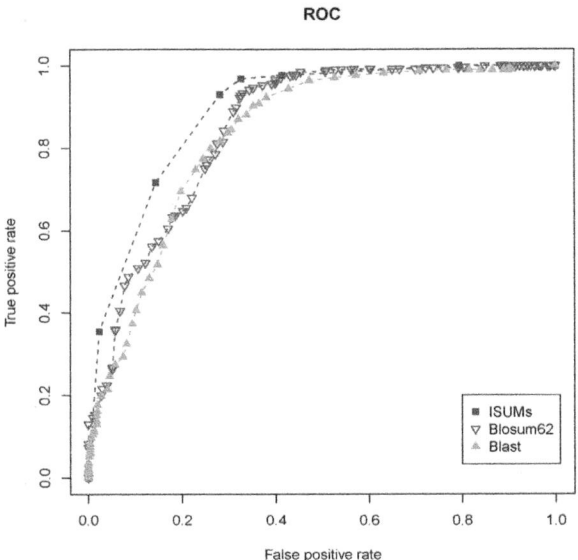

Fig. 3. ROC curve of Homeobox DNA motif predictions with ISUMs, BLOSUM62 and normalized BLAST E-values

3.6 Limitations of Sequence-Based DNA Motif Alignments

The set of pairwise alignments of DNA motifs, generated during the validation of ISUMs, were compared to the set of structure-based interface alignments of homeodomains (section 3.1, Figure 2). It was observed that sequence-based dynamic programming, as carried out by the STAMP software, occasionally yielded incorrect alignments, which failed to represent the underlying common interface architecture. These alignment errors occurred mainly when aligning DNA motifs of subtypes TAATnn and TGATAC. As can be seen in Supplementary Material, these difficult alignments were associated to an average STAMP E-value of 0.90.

4 Discussion and Conclusions

The results presented in this work support the hypothesis that the residues directly involved in DNA recognition can explain and capture binding preferences better than the complete protein sequence as aligned by BLASTP.

Work is under way with other families and so far the results obtained suggest that this holds true for other families, including bZIP and zinc finger transcription factors.

This work presents the performance of simple, binary ISUMs; it remains to be tested whether richer matrices, which would take longer to compute, can be created and whether different weights can be assigned to different interface positions to improve the observed DNA motif - interface correlation.

This protocol could in principle be used with different proteins families to establish a set of family-specific ISUMs that would help in the prediction of DNA motifs for *orphan* transcription factors. However, the kind of required data, such as the data produced by Noyes and collaborators, is unfortunately not available for most families. Nevertheless, our results provide quantitative evidence supporting the use of standard substitution matrices for evaluation of interface conservation, as previously suggested by other authors [16,17].

While our results support the general use of interface knowledge when evaluating sequence alignments of transcription factors, they also indicate that annotating interfaces can be particularly important in cases where full domain alignments yield poor scores, as in these cases highly similar interfaces can be masked by overall low similarity alignments.

It is important to recall that our results show that sequence-based alignment methods might fail to produce the correct DNA motif alignment between members of the same family, provided they are sufficiently divergent. This observation justifies the use of structural data for the comparison of transcription factors, whenever available, as done in this paper.

As a result of this work we have now added similarity scores and interface matrices to our weekly updated database 3D-footprint, which will make it easier to annotate and correctly align interfaces in different protein families.

Acknowledgements. This work was funded by Programa Euroinvestigación 2008 [EUI2008-03612].

References

1. Contreras-Moreira, B.: 3D-footprint: a database for the structural analysis of protein-DNA complexes. Nucleic Acids Res. 38(Database issue), D91–D97 (2010)
2. Berman, H.M., Westbrook, J., Feng, Z., Gilliland, G., Bhat, T.N., Weissig, H., Shindyalov, I.N., Bourne, P.E.: The Protein Data Bank. Nucleic Acids Res. 28(1), 235–242 (2000)
3. Murzin, A.G., Brenner, S.E., Hubbard, T., Chothia, C.: SCOP: a structural classification of proteins database for the investigation of sequences and structures. J. Mol. Biol. 247, 536–540 (1995)
4. Lewis, E.B.: A gene complex controlling segmentation in Drosophila. Nature 276, 565–570 (1978)
5. Noyes, M.B., Christensen, R.G., Wakabayashi, A., Stormo, G.D., Brodsky, M.H., Wolfe, S.A.: Analysis of homeodomain specificities allows the family-wide prediction of preferred recognition sites. Cell 133, 1277–1289 (2008)
6. Hertz, G.Z., Stormo, G.D.: Identifying DNA and protein patterns with statistically significant alignments of multiple sequences. Bioinformatics 15, 563–577 (1999)
7. Bailey, T.L., Williams, N., Misleh, C., Li, W.W.: MEME: discovering and analyzing DNA and protein sequence motifs. Nucleic Acids Res. 34, W369–W373 (2006)
8. Altschul, S.F., Gish, W., Miller, W., Myers, E.W., Lipman, D.J.: Basic local alignment search tool. J. Mol. Biol. 215, 403–410 (1990)
9. Mahony, S., Benos, P.V.: STAMP: a web tool for exploring DNA-binding motif similarities. Nucleic Acids Res. 35, W253–W258 (2007)

10. Contreras-Moreira, B., Sancho, J., Espinosa Angarica, V.: Comparison of DNA binding across protein superfamilies. Proteins 78(1), 52–62 (2009)
11. Edgar, R.C.: MUSCLE: multiple sequence alignment with high accuracy and high throughput. Nucleic Acids Res. 32, 1792–1797 (2004)
12. Henikoff, S., Henikoff, J.G.: Amino acid substitution matrices from protein blocks. Proc. Natl. Acad. Sci. 89, 10915–10919 (1992)
13. R Development Core Team: R: A Language and Environment for Statistical Computing. R Foundation for Statistical Computing, Vienna, Austria (2009)
14. Wingender, E., Dietze, P., Karas, H., Knuppel, R.: TRANSFAC: a database on transcription factors and their DNA binding sites. Nucleic Acids Res. 24, 238–241 (1996)
15. Berger, M.F., Badis, G., Gehrke, A.R., Talukder, S., Philippakis, A.A., Peña-Castillo, L., Alleyne, T.M., Mnaimneh, S., Botvinnik, O.B., Chan, E.T., Khalid, F., Zhang, W., Newburger, D., Jaeger, S.A., Morris, Q.D., Bulyk, M.L., Hughes, T.R.: Variation in Homeodomain DNA Binding Revealed by High-Resolution Analysis of Sequence Preferences. Cell 133(7), 1266–1276 (2008)
16. Luscombe, N.M., Thornton, J.M.: Protein-DNA interactions: amino acid conservation and the effects of mutations on binding specificity. J. Mol. Biol. 320(5), 991–1009 (2002)
17. Morozov, A.V., Siggia, E.D.: Connecting protein structure with predictions of regulatory sites. Proc. Natl. Acad. Sci. 104, 7068–7073 (2007)

Pyicos: A Flexible Tool Library
for Analyzing Protein-Nucleotide Interactions
with Mapped Reads from Deep Sequencing

Juan González-Vallinas[1,*,**], Sonja Althammer[1,*], and Eduardo Eyras[1,2]

[1] Computational Genomics Group, Universitat Pompeu Fabra, PRBB,
Dr. Aiguader 88 E08003, Barcelona, Spain
[2] Institució Catalana de Recerca i Estudis Avançats (ICREA), Passeig Lluís
Companys, 23, 08010 Barcelona, Spain
juanramon.gonzalezvallinas@upf.edu

Abstract. Deep DNA or RNA sequencing and posterior mapping to a
reference sequence is becoming a standard procedure in molecular biology research. Analyzing millions of mapped reads is a challenging task
that doesn't have a unique solution, because experiments using deep sequencing technology vary a great deal among each other. This is why we
have developed a flexible tool library called Pyicos, which aims to help
biologists in their research when performing their analysis on mapped
reads.

Keywords: Deep sequencing, High-throughput sequencing, ChIP-Seq,
CLIP-Seq, transcription factor, genomics, DNA, RNA, Peak Calling,
software development.

1   Introduction

Deep DNA and RNA sequencing methods are becoming a standard procedure
in molecular biology research. They are performed with commercial sequencing
machines [1], producing datasets of millions of short sequences. There are several
experimental protocols that are combined with deep sequencing to study Protein-
DNA and Protein-RNA interactions. For studying Protein-DNA interactions the
most used method is ChIP-Seq [2] . For the specific task of studying nucleosome
positioning, Mnase-seq [3] is being used. For Protein-RNA interactions the most
used methods are CLIP-Seq [4] and RIP-Seq.

Deep sequencing is overcoming some of the limitations that older techniques
have (Table 1) but it is also rising new computational challenges that need to be
addressed. One of this challenges is the efficient and correct analysis of mapped
reads. Telling the difference between real signal and noise is the main objective
of the methods developed for this task, commonly called "Peak callers" [8–10].

---

* These authors contributed equally to this work.
** Corresponding author.

A.T. Freitas and A. Navarro (Eds.): JBI 2010, LNBI 6620, pp. 83–88, 2012.
© Springer-Verlag Berlin Heidelberg 2012

Even experiments that share the same protocol can produce divergent types of mapped datasets that need to be analyzed with different approaches. This is the case of ChIP-Seq: Experiments targeting transcription factors normally produce short narrow clusters of "punctuated" data, usually covering few hundreds of bases per cluster. Alternativerly, if the target is a histone modification, the clusters will span several hundreds of kilobases [11]. In CLIP-Seq experiments, if the target protein is binding to pre-mRNA it will be necessary to map the reads to the genome, but if it binds mainly to mRNA, the mapping will be better done directly to the transcriptome.

1.1   Motivation

The mentioned "Peak Callers" methods are generally designed for a specific type of experiment and a specific type of data. In our lab we deal with an increasing amount of heterogeneous data, this is why we developed Pyicos, a flexible tool library for the analysis of short reads. We aimed to analyze data from different ChIP-seq and CLIP-seq datasets coming from the Solexa/Illumina platform with a common tool library.

2   Results

2.1   General Features

Modularity. Pyicos modular architecture provides easy addition and modification of functionality, like new input or output formats, artifact detection conditions, pre-processing and post-processing operations, interfaces, etc.

Memory Usage. Pyicos avoids loading whole datasets in memory and its performance is not dependant on it. Mapped reads datasets can occupy up to several gigabytes in disk and presumably will get bigger as sequencing machines improve, so this feature is critical for the future performance of the tool.

Table 1. Comparison of different experimental methods ChIP-Chip signal-to-noise ratio is very high and its resolution is limited to the size of the probes (~200bp) [12]. Sanger sequencing methods can take up to a year to sequence 1 gigabase at a cost of $0.10 per 1000 bases whereas deep sequencing with Illumina produces 1 gigabase in half a day with a cost of $0.001 per 1000 bases.

Classic Method	New Method	Use	Advantages
Sanger Sequencing	Mnase-Seq	Nucleosome Positioning	Genome wide, cheaper
ChIP-Chip	ChIP-Seq	protein-DNA interactions	Specificity, sensitivity, genome wide
Sanger Sequencing	CLIP-Seq RIP-Seq	protein-RNA proteins interactions	Genome wide, cheaper

Command Line Interface. Pyicos can be used as a Unix-like tool through its command line interface. The technical details of the usage can be read directly from the interactive help menu by typing in the console:

```
>pyicos --help
```

Pyicos command line tool is composed by a set of sub-commands:

```
>pyicos sub-command --help
```

Protocol File. Alternatively, a protocol file can be provided to the software, indicating the operations to be performed.

2.2   Operations

All mapped read analysis software apply a combination of heuristics and statistical operations to discern the real signal from the noise. We identified the operations we needed and "atomized" them, allowing them to be used independently or combined together in a pipeline. The following are a few examples of implemented operations in Pyicos.

Strand Cross-correlation and Extension. When preparing the samples of ChIP and CLIP for the sequencing machines, a mean read length is selected from the agarose gel corresponding to the expected size of the DNA or RNA bound to the protein. However, the length of the sequenced fragments may not be as expected because Solexa/Illumina solid-state PCR favors shorter sequences [11], or it simply may not be known. In this case, Pyicos can perform a cross-correlation test between forward and reverse strand clusters [10] in order to find the optimal extension length. The strcorr operation provides a correlation plot and the extension length with the best correlation score. The extend operation can use the value provided by the cross-correlation test, or alternatively, it allows a manual insertion of an extension value.

Normalize. When comparing two or more datasets of mapped reads to each other, normalizing is a necessary step. Pyicos calculates a normalization factor with the sum of all reads lengths in the control, and dividing them by the sum of all read lengths in experiment.

Subtract. Control datasets are normally obtained by sequencing a sample with a unspecific antibody, like IgG. When a control dataset is provided, Pyicos gives the possibility to subtract the signal of the control to the sample signal in order to obtained enriched regions. Subtract is a novel algorithm that we introduce with Pyicos. Pyicos subtracts with nucleotide precision one dataset from another, unlike other methods like Findpeaks[8] or MACS [9], which use statistical approximations for the subtraction.

Split (find "sub-clusters"). If two signal events occur very close together there is a possibility that their reads will overlap, being mistaken as one cluster instead of two and generating a "double cluster" shape. The split operation can then be used to recognize these patterns and separate overlapping signals. This operation is useful in punctuated TF data, but also interesting for broad data when looking for enrichment in nucleosome positioning.

Remove Regions. Some regions can be a source of noise due to contamination of the experiment, an error in the sequencing or an error in the mapping. Using a region "black list" to exclude these regions from the analysis is a convenient pre-processing step to overcome these problems and improve the outcome of the experiment.

Significance with Poisson Analysis. We designed 3 tests based on the Poisson distribution to determine the threshold for filtering and trimming the clusters, classifying them by p-value. The accumulated p-values for all 3 tests are calculated with the Poisson density function, where both x and λ are defined per test.

In the "Max Height" test, x is the height of the cluster and λ is the average height of clusters in a given region. This test will favor clusters that have greater heights. This is the default test for the `filter` operation.

In the "Number of Reads" test, x is the number of reads in the cluster and λ is the average number of reads per cluster in a given region. Clusters with a large quantity of reads will be considered more significant, independently of their height.

In the Reads per nucleotide test, x is the coverage of reads in a nucleotide position and λ is the average reads per nucleotide inside the clusters in a given region.

Filter Clusters According to a Threshold. After deciding what criteria to use for filtering and classifying the dataset, the filter operation must be applied. Pyicos `filter` operation uses "Poisson Max Height" analysis by default with a p-value of 0.01. Filtering can also be used as an operation independent from the statistical calculation if needed, using the flag `--force-threshold`.

Modified FDR. This variation of the FDR method was introduced for the analysis of FOX2 splicing regulator [13] CLIP-Seq experiment. This method is used when we do not have a control. We randomize the positions of the sample reads in a given region, keeping the length and the number of reads, generating for each position r a randomized coverage n_r. For each density of reads h, we define a randomized probability $P_R(h)$:

$$P_R(h) = \frac{1}{N} \sum_{r \geq h} n_r$$

We repeat this process k times (by default $k = 100$) , calculating a value $P_R(h, j)$ for the jth iteration. When all iterations are complete we calculate the mean $\mu_R(h)$ and standard deviation $\sigma_R(h)$ for all the random probabilities $P_R(h, j)$. Finally, we can calculate the modified FDR for a given density of reads h:

$$\widetilde{FDR}(h) = \frac{\mu_R(h) + \sigma_R(h)}{P(h)}$$

Then for each density h we can check if $\widetilde{FDR}(h) < t$ where t is the given False Discovery Rate limit, by default 0.01.

3   Discussion

We developed Pyicos, a novel tool and library that has a flexible approach to the analysis of short mapped reads. The division of the software into "operations" allows a greater customization and control of the analyses carried out and provides applicability across different experiment types. The combination of this flexibility, efficient memory performance, modular architecture and novel algorithms for the threshold definition and cluster subtraction gives us unique advantages and increases the possibilities of usage of the software.

Availability

The software is released with an open-source GPL license. The source code for download and a more technical online documentation are available at: http://regulatorygenomics.upf.edu/pyicos

References

1. Bennett, S.: Solexa Ltd. Pharmacogenomics 5(4), 433–438 (2004)
2. Robertson, G., Hirst, M., Bainbridge, M., Bilenky, M., Zhao, Y., Zeng, T., et al.: Genome-wide profiles of STAT1 DNA association using chromatin immunoprecipitation and massively parallel sequencing. Nat. Methods 4(8), 651–657 (2007)
3. Schones, D.E., Cui, K., Cuddapah, S., Roh, T., Barski, A., Wang, Z., et al.: Dynamic regulation of nucleosome positioning in the human genome. Cell 132(5), 887–898 (2008)
4. Sanford, J.R., Wang, X., Mort, M., Vanduyn, N., Cooper, D.N., Mooney, S.D., et al.: Splicing factor SFRS1 recognizes a functionally diverse landscape of RNA transcripts. Genome Res. 19(3), 381–394 (2009)
5. Trapnell, C., Salzberg, S.L.: How to map billions of short reads onto genomes. Nat. Biotechnol. 27(5), 455–457 (2009)
6. Li, H., Ruan, J., Durbin, R.: Mapping short DNA sequencing reads and calling variants using mapping quality scores. Genome Research 18(11), 1851–1858 (2008)
7. Li, R., Li, Y., Kristiansen, K., Wang, J.: SOAP: short oligonucleotide alignment program. Bioinformatics 24(5), 713–714 (2008)
8. Fejes, A.P., Robertson, G., Bilenky, M., Varhol, R., Bainbridge, M., Jones, S.J.M.: FindPeaks 3.1: a tool for identifying areas of enrichment from massively parallel short-read sequencing technology. Bioinformatics 24(15), 1729–1730 (2008)
9. Zhang, Y., Liu, T., Meyer, C.A., Eeckhoute, J., Johnson, D.S., Bernstein, B.E., et al.: Model-based analysis of ChIP-Seq (MACS). Genome Biol. 9(9), R137 (2008)
10. Kharchenko, P.V., Tolstorukov, M.Y., Park, P.: Design and analysis of ChIP-seq experiments for DNA-binding proteins. Nat. Biotechnol. 26(12), 1351–1359 (2008)

11. Pepke, S., Wold, B., Mortazavi, A.: Computation for ChIP-seq and RNA-seq studies. Nat. Meth. 6(11s), S22–S32 (2009)
12. Gottardo, R.: Modeling and Analysis of ChIP-Chip Experiments (Internet). In: Chromatin Immunoprecipitation Assays, pp. 133–143 (2009), http://dx.doi.org/10.1007/978-1-60327-414-2_9
13. Yeo, G.W., Coufal, N.G., Liang, T.Y., Peng, G.E., Fu, X., Gage, F.: An RNA code for the FOX2 splicing regulator revealed by mapping RNA-protein interactions in stem cells. Nat. Struct. Mol. Biol. 16(2), 130–137 (2009)

Novel Angiogenic Functional Targets Predicted through *"Dark Matter"* Assessment in Protein Networks

Ian Morilla[1], Miguel A. Medina[1,2], and Juan A.G. Ranea[1,2]

[1] Department of Molecular Biology and Biochemistry,
University of Malaga, Campus de Teatinos s/n, 29071, Malaga, Spain
[2] CIBER de Enfermedades Raras (CIBERER), Valencia, Spain

Abstract. In order to model protein networks we must extend our knowledge of the protein associations occurring in molecular systems and their functional relationships. We have significantly increased the accuracy of protein association predictions by the meta-statistical integration of three computational methods specifically designed for eukaryotic proteomes. From this former work it was discovered that high-throughput experimental assays seem to perform biased screenings of the real protein networks and leave important areas poorly characterized. This finding supports the convenience to combine computational prediction approaches to model protein interaction networks. We address in this work the challenge of integrating context information, present in predicted and known protein network models, to functionally characterize novel proteins. We applied a random walk-with-restart kernel to our models aiming at fixing some poorly described or unknown proteins involve in angiogenesis. This approach reveals some novel key angiogenic components within the human interactome.

Keywords: protein domain fusions, ppi networks, functional prediction proteins, networks topology, genes candidate prioritization and random walks with restart.

1 Introduction

In the current work it is shown how exploiting the context information present in protein-protein networks we achieve to predict novel candidates to be involved in Angiogenesis process which is important in tumoral growing.

Predicted protein network models were built based on the integration of different computational methods, which exploit diverse and orthogonal information, such as domain co-occurrence analysis, gene expression similarity and interacting proteins homology signals. We assessed the potential of this bio-computational approach to obtain accurate and comprehensive protein interactions network models of the human proteome and compare these to protein interactome models of the same species based on the combination of all the experimentally inferred protein association from seven comprehensive biological databases. The protein

A.T. Freitas and A. Navarro (Eds.): JBI 2010, LNBI 6620, pp. 89–93, 2012.

network model based on predictions was termed Predictiogram (PG), while the network modeled with experimental data was named Knowledgegram (KG).

Complex network models hoard context information about novel functional associations beyond the simple binary protein interactions. In order to exploit this meta-information we used one mathematical tool that enables us to functionally characterize proteins by integrating all the interactions defined in network models related to a given functional context. This strategy allows weighting each protein relationship to any pre-determined functional group in a network model.

We implement a random walk-with-restart (RWR) algorithm for generating ranked lists of gene candidates related to the human angiogenic process [1]. RWR is a powerful network distance measure that avoids local topological artifacts relying on highly connected protein or hubs.

2   Results, Conclusions and Methods

2.1   Results

We assessed the random walk-with-restart (RWR) method performance for ranking proteins in the angiogenic dataset using the leave on out method benchmarking with the PG and KG network models (Fig. 1). The AUC (area under the curve) values of the receiver-operating characteristic (ROC) curves validation plots were calculated for the angiogenic set and the two implemented random models (Fig. 1 panels A and B). AUC values comparison shows that the RWR re-annotation method significantly outperfoms random ranking for re-annotating angiogenic proteins.

Position in the RWR prioritization ranked lists allows assigning an angiogenesis functional probability to each protein in the reference interactome. The numerical results show that the confidence interval for these predicted probabilities is slightly better in the PG than in KG model (see Predicto and Knowledgegrams construction), but in both cases the performance is remarkable. Notice that in angiogenesis about 80% of the query dataset are above the 30% top ranked proteins list (Fig. 1, panel B).

The plots in **Fig. 1** correspond to cumulative curves of the ranked angiogenic query protein sets and were generated for the KG (panel A) and PG (panel B) protein network models (89 and 59 prioritizations respectively). Normalised axis show sensitivity versus ranking in average. Two random models (coloured in red and green in Fig. 1 -see Methods-) reflect the lack of randomness in our protein models KG and PG. A perfect classifier would have an AUC of 100% and a random one would have an AUC of 50%.

The AUC value calculated after applying leave one out cross validation based on RWR algorithm did not present any remarkable difference when the restart probability γ (γ is a parameter used in the base ranker formula see Methods) was shuffled from 0.5 to 0.9. Although those best performance for γ was at 0.7. These consistent results with different γ values indicate the robustness and stability of the outcome in the applied methodology.

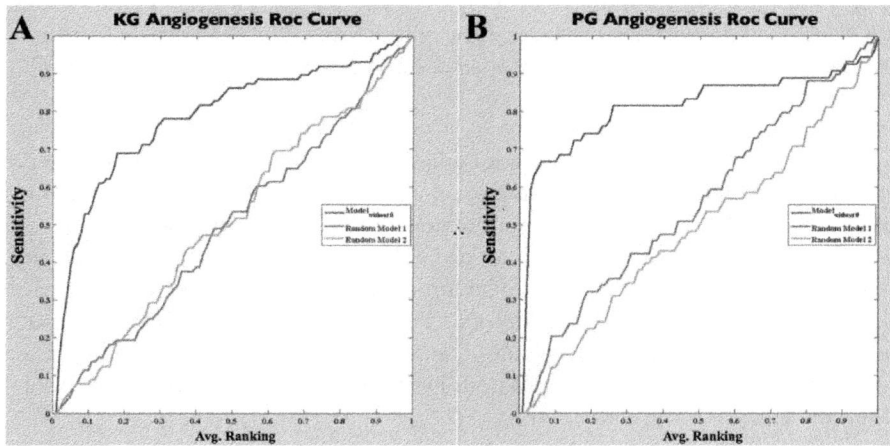

Fig. 1. Benchmarking Results

2.2   Conclusions

RWR has revealed itself as a powerful mathematical method to integrate interaction information present in protein network models to predict the most probable functional context of single proteins. RWR, like other methods, which measure distances between nodes in modeled interactomes, enables us passing from protein-protein interaction information to ranked list of proteins, by weighting each protein probability to belong to a pre-determined functional group. The shown methodology is useful as to functionally annotate unknown proteins as to get increasing the experimental efficiency in the characterization of new components of biological systems by prioritizing the most promising targets.

2.3   Methods

2.3.1   Predictogram (PG) and Knowledgegram (KG) Construction
Human PG protein network model for human proteome based on Näives-Bayes integration was obtain from [2] Supp. Material. For the PG were selected the protein interactions with p-value corresponding with precision $\geq 90\%$. the KG includes the protein-protein association data from Reactome, Kegg, Intact, MINT and HRPD databases.

2.3.2   Angiogenesis Dataset Source
A dataset of 116 proteins related to angiogenesis process was gathered for this work from manually curation of literature and biological databases.

2.3.3   Random Walk-with-restart Kernel
By definition, a *kernel* [3] is a function that gives the dot product between 2 vectors in some multi-dimensional space (called feature space). A kernel matrix (often abbreviated as kernel) contains the evaluation of the kernel function for

all pairs of data points under consideration. In this way, a kernel can be viewed as a matrix of similarities between data points and different kernels capture different notions of similarity as they correspond to embedding the data in different feature spaces.

A *random walk* [1] simulates a random walker that starts on a source node i or a set of them simultaneously. At each step, the walker selects randomly among its immediate neighbours and jumps to a node j (based on edge weights in accordance with the following probability $p_{ij} = P(s(t+1) = j \mid s(t) = i) = a_{ij}/a_{i\bullet})$. A random walk is a type of Markov chain from which different measures of similarity between nodes of the graph can be computed. Specifically the random walk with restart (RWR) [1] (where the walker has a small probability at each step to start again from a random node) exploits the above kernel methodology. Formally, this method is defined by:

$$\begin{cases} x(0) = e_i \\ x(t+1) = \alpha P^T x(t) + (1-\alpha)e_i \end{cases} \tag{1}$$

2.3.4   Performance Measurement

In order to measure the performance of the whole optimization and training procedure, leave-one-out cross-validation [4] which is illustrated in Figure 2 was used for each query (those proteins related to spindle either angiogenesis) protein in the *entire human interactome*, unlike [4] where a nearest genomic interval to the query genes is pre-selected.

Therefore, for each spindle or angiogenesis gene, we obtain a rank list of test genes, that is, prioritizations of the whole proteins implied in spindle or angiogenesis. By following this reasoning, the held-out should be nearby the top of the remaining human interactome proteins tested. Obviously, nearer this query held-out protein is better performance it is obtained by RWR. We stress our methodology with a biological process, i.e., the angiogenesis process in two different models; KG and PG (see KG and PG construction). So once the cross validation was carried out, we obtained 2 ranking lists, angiogenesis with 89 prioritizations (89 angiogenic proteins mapped out 116) for the KG and other with 59 prioritizations out 116 possible for the PG. For those, we work out sensitivity and the average of the tested set of proteins (the entire interactome once held out the corresponding query protein) at shuffling thresholds. Sensitivity is defined as the percentage of query proteins that were ranked above a particular threshold. Average of the tested set is defined as an increasing sequence of percentages. So, if a sensitivity/avg. value of 80/40 means that the 80% of the query proteins is located above the top ranked 40%. Finally, we present the results in normalised axes (x and y -axis- values $\in [0,1]$ in $\mathbb{R}^+$) ROC curves plots and use the area under the curve (AUC) for a better understanding of the measure. Here, wider area under the curve better ranked every held-out protein is. In order to avoid any suspicious of randomness we performed two different random models

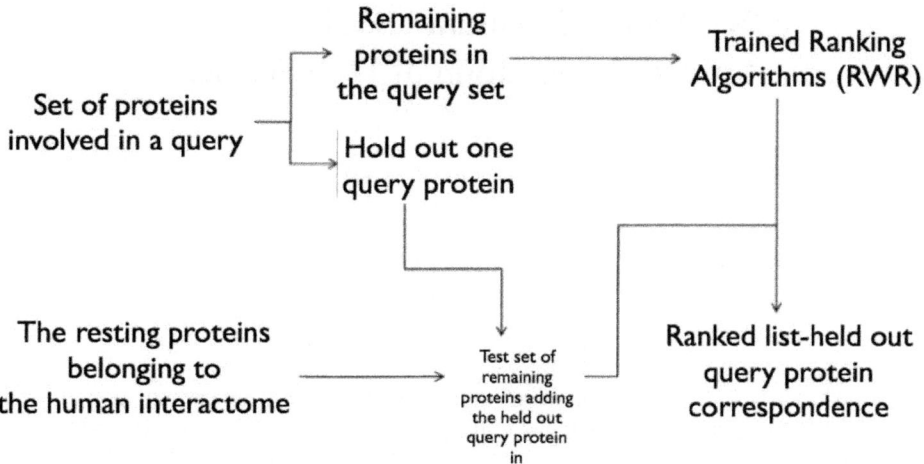

Fig. 2. Leave one out validation scheme

as shown in panels A and B, from the Fig. 1. Random Model 1 (coloured in red in Fig. 1) corresponds to randomly selecting one gene from the genome as target instead of a gene from the pathway. And Random Model 2 (coloured in green in Fig. 1) corresponds to a query set randomly selected and with the same size of the source query/pathway dataset.

References

[1] Köhler, S., Bauer, S., Horn, D., Robinson, P.N.: Walking the Interactome for Prioritization of Candidate Disease Genes. The American Society of Human Genetics (2008), doi:10.1016/j.ajhg.2008.02.013
[2] Ranea, J.A.G., Morilla, I., Lees, J.G., Reid, A.J., Yeats, C., Clegg, A.B., Sanchez-Jimenez, F., Orengo, C.: Finding the "Dark Matter" in Human and Yeast Protein Network Prediction and Modelling. PLoS Comput. Biol. 6(9), e1000945 (2010), doi:10.1371/journal.pcbi.1000945
[3] Fouss, F., Franoisse, K., Yen, L., Pirotte, A., Saerens, M.: An Experimental Investigation of Graph Kernels on Collaborative Recommendation and Semisupervised Classification. In: Proceedings of the Eighth International Conference on Data Mining, ICDM 2009 (2009)
[4] Li, Y., Patra, J.C.: Integration of multiple data sources to prioritize candidate genes using discounted rating system. BMC Bioinformatics 11(suppl. 1), S20 (2010), doi:10.1186/1471-2105-11-S1-S20

jORCA and Magallanes Sailing Together towards Integration of Web Services

Johan Karlsson[1] and Oswaldo Trelles[2]

[1] Fundación IMABIS, Málaga, Spain
[2] Computer Architecture Department, University of Málaga, 29071 Málaga, Spain
{tjkarlsson,ortrelles}@uma.es

Abstract. It becomes increasingly important to support automated service discovering and composition due to the growing number of Web Services and data types in bioinformatics and biomedicine. jORCA is a user-friendly desktop client which is able to discover and invoke Web Services from different metadata repositories for services. This paper demonstrates the usefulness of jORCA for service composition by recreating a previously published workflow, starting with the discovery of data types, service composition (workflow generation) and refinement; to enactment, monitoring and visualization of results. The system has been exhaustively tested and documented and is freely available at http://www.bitlab-es.com/jorca.

1   Introduction

Bioinformatics is strongly based on the Web to deploy tools and provide universal access to distributed resources. However, a weakness of using Web tools –in the general sense of applications available on the Web- is that they differ in their descriptions, invocation protocols and data formats. This represents a strong barrier to interoperability and discovery of web tools. Web Services (WS) has appeared as a potential technology which addresses these problems. Several clients have been developed to assist in the utilization of WS in bioinformatics, see for example Gbrowse [15], MOWServ [9] and Seahawk [2]. Although these clients have facilitated WS usage, they have typically been developed and designed for use with a certain WS technology.

To facilitate seamless integration of WS in bioinformatics, we developed jORCA [6] which is a desktop client operating with a uniform representation of different web resources enabling advanced WS integration. In the same line, we developed Magallanes [11] which addresses two important activities:

- WS *discovery*, which extends traditional syntactic text-search engines and filtering mechanisms with inexact matches based on a scoring system. Additionally, it can provide 'did you mean…?' suggestions to guide users in constructing search queries.
- WS *composition*, which involves connecting compatible WS as pipelines (simple workflows). Pipelines can be automatically generated by

A.T. Freitas and A. Navarro (Eds.): JBI 2010, LNBI 6620, pp. 94–101, 2012.
© Springer-Verlag Berlin Heidelberg 2012

selecting input and output data types. Note that those data types can be discovered through Magallanes. If there are multiple options, users are requested to select the more suitable option.

Magallanes has been fully integrated in jORCA. Recent extensions to jORCA related to workflows were reported in [3] and included workflow enactment and monitoring within jORCA as well as extended support for workflow composition by means of drag and drop from the history of WS invocations.

The purpose of this paper is to describe a case study which demonstrates the utility of using jORCA to create potentially interesting workflows. The study reproduces a previously published workflow [5] for clustering of co-regulated genes to obtain a hierarchical multiple sequence alignment based on similarity of promoter configurations and thereby displaying the most relevant common features of those genes at sequence level.

jORCA architecture is extensively described in [6]. In summary, the software combines different specialized software components. WS metadata is represented uniformly using MAPI [10] and currently components exist to connect jORCA to BioMOBY WS registries, a registry for biomedical grid-services (ACGT project [12]) and for WSDL described WS (however with limited support for data types). In this paper, we have focused on using BioMOBY which has an active developer community and almost 1600 available bioinformatics WS in the main registry [7] (beginning of 2009). By using BioMOBY, we assure WS compatibility whenever semantically reasonable and thereby facilitate their composition.

The organization of this paper is as follows: in this section, we introduce jORCA and Magallanes. We proceed to re-create the workflow from [5] using these tools in Section 2. Different aspects of the integration of Magallanes in jORCA are discussed in Section 3. Finally, Section 4 concludes the paper.

2   Service Composition Using jORCA

In [3], we described new functionality in jORCA, including improved WS composition and enactment of the resulting workflows. In order to better demonstrate the new functionality, we will reproduce the workflow proposed in [5] by using the automatic WS composition of Magallanes.

In this scenario, a researcher has obtained a set of sequences in FASTA format. To locate an initial data type, Magallanes can be used to search for a datatype with the search string "nueclid acid sequence". Notice that the spelling of "nucleic" was incorrect but Magallanes suggest the correct spelling "nucleic acid sequence". The datatype FASTA_NA_multi is selected by the researcher after reading the data type descriptions.

The goal of the exercise (i.e. the output) is to find a Newick representation of the hierarchical multiple sequence alignment. Searching for "newick" with Magallanes locates the data type Newick_Text.

It is now possible to use the automatic WS composition of Magallanes using the input and output data types, resulting in the automatically created pipeline illustrated in Figure 1.

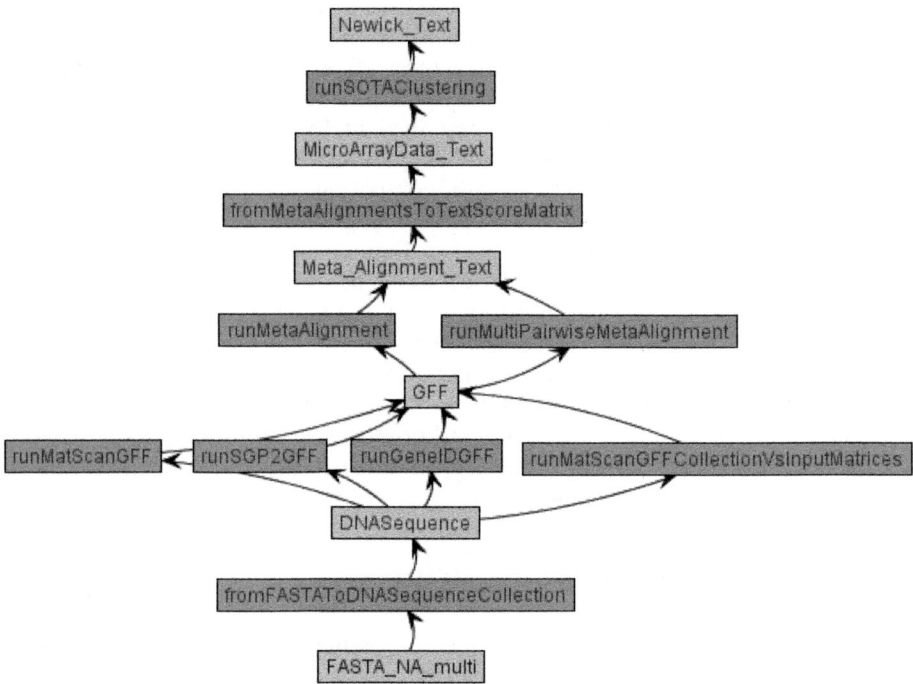

Fig. 1. Initial workflow created by Magallanes. Note the different options which have to be manually reviewed by the workflow designer.

Several options are possible to go from a DNASequence to GFF (Generic Feature Format) and from GFF to a meta alignment. The user reads the service descriptions by moving the mouse pointer over the options and finally selects the runMatScanGFF and RunMultiPairwiseMetaAlignment which results in the final pipeline visualized in Figure 2.

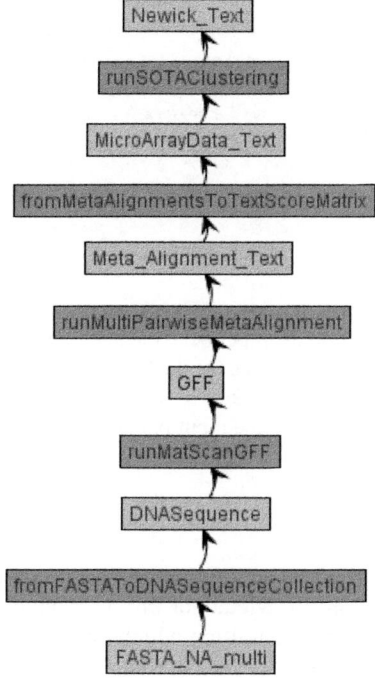

Fig. 2. Final pipeline created after manual selection by workflow designer

This pipeline can be exported as a Taverna compatible workflow or directly enacted within jORCA. We will choose the second option, which brings up the user interface in Figure 3.

When enacting a pipeline within jORCA, the required parameters are requested and marked with special colors. By default, only the parameters of the first WS in the pipeline and the suggested names of the outputs are displayed by jORCA. The intermediate WS use the default parameters (if any). These parameters are hidden to avoid a complex interface but the user can choose to edit by clicking on the WS.

During enactment, jORCA allows the user to monitor the progress by colouring the WS in the graph (see Figure 3). Intermediate results can also be saved in the user file system where they can be visualized or used for input for other WS/pipelines, just as the final output of the pipeline. In this example, external software (i.e. a Newick tree viewer) can be used from within jORCA to display the result.

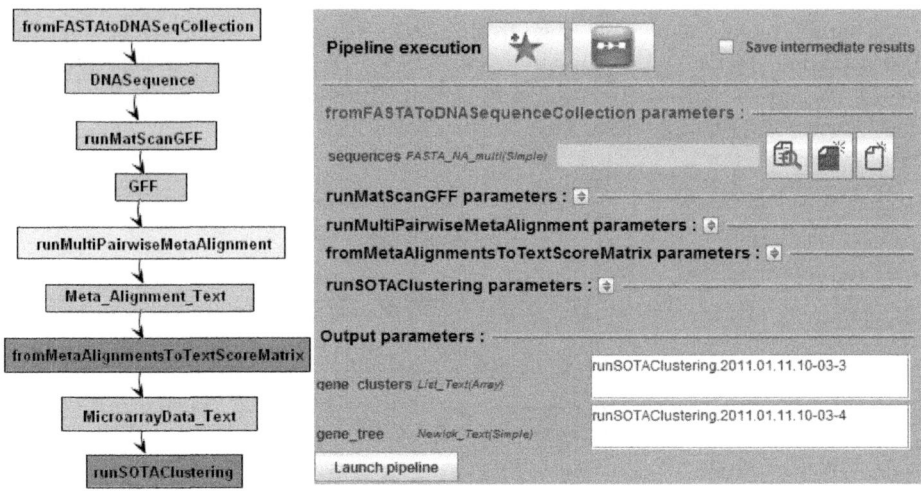

Fig. 3. This figure shows the monitoring of workflows (pipelines) enactments in jORCA. The top part shows clickable icons to add the pipeline to the list of favorites (i.e. save the pipeline for later use) or exporting the pipeline to Taverna. On the left side, the progress of the enactment is shown. Successfully invoked WS will be marked in green, running invocations in yellow and failed invocations in red. On the right side, the input parameters of the different WS are displayed.

3   Discussion

Naturally, the quality of service annotations is fundamental for the feasibility of automatic service composition. Metadata in the service repository that we use [8] is pre-approved by an ontology committee which ensures the quality of the metadata.

The main BioMOBY registry [7] maintains an open policy for registering datatypes and services. While this is flexible and allows developers to quickly publish their metadata, some issues have arisen due to this openness which complicates service composition. One such issue is the lack of re-use of existing data types when developers simply register a new, unique data type for their service without checking if a suitable data type already exists. Effective WS composition is only feasible if WS share data types for their inputs/outputs to a high degree.

Another issue is that all sequence retrieval services declare their input data type as the base class of the ontology. Users are, by convention, supposed to enter sequence ID and data source (for example UniProt) in their inputs to those services. Nevertheless, this affects the service composition of Magallanes if the user selects to use inheritance because those services, by definition, are compatible with any input data. However, for example, sending a sequence to a sequence retrieval service does not make much sense.

It is clear that great care has to be taken with the ontologies. The approach taken to review new registrations before making them public by means of a committee is therefore useful. Community activities ("jamborees") could also, in general, be a way to agree, discuss and improve service annotations.

The original workflow reported in [5] was developed manually. It is difficult to compare the effort required to construct that workflow with the one developed in Section 2 since [5] also included the development of some new WS. However, for a typical researcher, manually constructing such a workflow would require time-consuming review of available web-services and data types. While most high-level tools for using BioMOBY WS include functionality to find compatible WS (based on the data type hierarchy), automatic composition of WS is rare but very useful as the example provided in Section 2 shows.

The approach in [1] is an interesting example of interactive service composition where only the user is guided in each step by displaying only compatible and potentially useful services (measured by a combination of semantic input/output similarity and a rough estimation of WS popularity). The approach in this paper is less interactive and instead focuses in automatically generating a good initial pipeline and letting users either directly enact or further refine the pipeline in advanced third-party tools such as Taverna. We also address discovery of the input/output data types for the pipelines, something not addressed in [1].

Taverna also has a plug-in for BioMOBY services [4] but is intended to be used as a base for constructing workflows inside Taverna. The plug-in does not automate in any way the service composition or facilitate service discovery in the way Magallanes does.

3.1 Integration of Magallanes in jORCA

There are two ways to use Magallanes in external applications: directly using the provided graphical user-interface (UI) or create a custom GUI. In jORCA, we used the provided UI. The integration was straightforward since both jORCA and Magallanes use MAPI as a base. There is support for plug-ins in jORCA and using that mechanism we provided an instance of the currently loaded service repository to the Magallanes GUI and registered jORCA to listen to Magallanes GUI events.

The alternative approach, to directly use the Magallanes programming library and create our own GUI was possible but not needed since the existing GUI was visually compatible with the rest of jORCA. However, other applications with different GUIs can take advantage of the search library to create customized searches specifying, for example, if keywords should be case-sensitive or combined with AND, OR or even regular expressions. The search library also provides functionality to access Magallanes "did you mean" suggestions which are sorted according to relevancy (based on the similarity of descriptions to the search query and history of earlier user selections, see [11] for details). Service composition is also supported in the Magallanes programming library.

Magallanes is also used in the ACGT architecture for service discovery where the GUI is based on a special kind of web-pages, technically very different from the stand-alone user interface of jORCA.

4   Conclusions

Although the diverse set of web tools in bioinformatics is fundamentally an important and positive resource, the profusion and dispersion of bioinformatics tools make routine tasks of data processing complicated. End-users need to perform tedious and error-prone copying of results from one web tool to another. Even more, it is often necessary to adjust the format or data for further processing with other tools. Workflows and WS are useful solutions to simplify repetitive and complex data processing in bioinformatics.

In previous communications, we reported two different applications (jORCA and Magallanes) to facilitate the discovering, interconnection and exploitation of bioinformatics tools.

Magallanes is equipped with a powerful search engine specialized in the discovery of WS by exploring all the available metadata descriptions of the resources through syntactic and regular-expressions search methods. It also includes well-known and popular methods to fix misspelled keywords and a feedback module to learn from the user preferences. Additionally, it is also able to automatically connect WS into pipelines (simple workflows).

jORCA is aimed at facilitating the invocation of WSs by exploiting both the uniform representation of WS provided by MAPI and the discovery and composition of workflows supplied by Magallanes. Therefore, it is able to discover resources, request parameters from the user in a uniform way, invoke services and workflows with independence of their deployment style, monitor the service execution progress and visualize the results. All these functionalities are included in a modular and robust application with a user friendly design for a broad range of user skills. Several features of jORCA are designed to simplify usage of WSs, including drag and drop methods, intelligent data auto-conversion, advanced features such as service mirroring, manual and automatic invocation scheduling, asynchronous calls, execution tracing and logbook to redo repetitive tasks more easily.

The utility of the workflow functionality of jORCA and Magallanes is demonstrated in this paper by reproducing a previously published workflow in an effective and intuitive way, including monitoring of enactment and visualizing the end-results.

Future planned improvements include functionality to suspend and resume workflow enactment, improved editing of workflows, including complex workflows (multiple branches). We also plan to provide better support for calling locally installed software to process data (currently possible but complex for the typical end-user).

Acknowledgements. This work has been partially financed by the National Institute for Bioinformatics (www.inab.org) a platform of ISCIII; the EU project "Advancing Clinico-Genomic Trials on Cancer" (EU-contract no. 026996) and the RIRAAF Spanish network on allergies (RD07/0064/0017).

References

1. DiBernardo, M., Pottinger, R., Wilkinson, M.: Semi-automatic web service composition for the life sciences using the BioMoby semantic web framework. Journal of Biomedical Informatics 41, 837–847 (2008)
2. Gordon, P.M.K., Sensen, C.W.: Seahawk: Moving Beyond HTML in Web-based Bioinformatics Analysis. BMC Bioinformatics 8, 208 (2007)

3. Karlsson, J., Martín-Requena, V., Ríos, J., Trelles, O.: Workflow Composition and Enactment Using jORCA. In: Margaria, T., Steffen, B. (eds.) ISoLA 2010. LNCS, vol. 6415, pp. 328–339. Springer, Heidelberg (2010), doi:10.1007/978-3-642-16558-0_28
4. Kawas, E., et al.: BioMoby extensions to the Taverna workflow management and enactment software. BMC Bioinformatics 7, 523 (2006)
5. Kerhornou, A., Guigó, R.: BioMoby Web Services to support clustering of co-regulated genes based on similarity of promoter configurations. Bioinformatics 23(14), 1831–1833 (2007)
6. Martín-Requena, V., Ríos, J., García, M., Ramírez, S., Trelles, O.: Jorca: easily integrating bioinformatics Web Services. Bioinformatics 26(4), 553–559 (2010)
7. MOBY Central at the University of Calgary, http://moby.ucalgary.ca/moby/MOBY-Central.pl
8. MOBY Central at University of Malaga, http://chirimoyo.ac.uma.es/cgi-bin/MOBY/MOBY-Central.pl
9. Navas-Delgado, I., et al.: Intelligent client for integrating bioinformatics services. Bioinformatics 22, 106–111 (2006)
10. Ramirez, S., et al.: A flexible framework for the design of knowledge-discovery clients. In: International Conference on Telecommunications and Multimedia (2008)
11. Ríos, J., Karlsson, J., Trelles, O.: Magallanes: a Web Services discovery and automatic workflow composition tool. BMC Bioinformatics 10, 334 (2009)
12. Tsiknakis, M., et al.: Building a European biomedical grid on cancer: the ACGT Integrated Project. Studies in Health Technology and Informatics 120, 247 (2006)
13. Web services architecture working group (September 2006), http://www.w3.org/2002/ws/arch/
14. Wilkinson, M.D., et al.: Interoperability with Moby 1.0–it's better than sharing your toothbrush! Briefing in Bioinformatics 9(3), 220–231 (2008)
15. Wilkinson, M.D.: Gbrowse moby: a web-based browser for BioMOBY services. Source Code for Biology and Medicine 1, 4 (2006)

iHOP Web Services Family

José M. Fernández[1], Robert Hoffmann[3], and Alfonso Valencia[2]

[1] GN2, Spanish National Bioinformatics Institute (INB)
[2] Spanish National Bioinformatics Institute (INB)
Structural Biology and Biocomputing Programme, CNIO, Spain
{jmfernandez,avalencia}@cnio.es
[3] Computational Biology Center,
Memorial Sloan Kettering Cancer Center, New York NY 10065, USA
hoffmann@cbio.mskcc.org

Abstract. iHOP provides fast, accurate, comprehensive, and up-to-date summary information on thousands of biological molecules by automatically extracting key sentences from millions of PubMed documents. iHOP web services are providing public programmatic access to all this information since their publication in 2007.

This manuscript describes recent improvements on the iHOP web services family and some of the scenarios in which the web services have been applied.

Availability. iHOP web services family is documented at its website http://ws.bioinfo.cnio.es/iHOP/

Keywords: Text mining, web services, whole genome analysis.

1 Introduction

The iHOP[9] literature mining server allows researchers to explore a network of gene and protein interactions by directly navigating the public set of scientific manuscripts where they are co-mentioned. iHOP web services were made publicly available in 2007[7]. At that time there were around 80 000 biological molecules indexed by the iHOP literature server, mainly from a few selected model organisms. Currently, iHOP handles more than 6 000 000 biological molecules, a situation which has expanded the number of scenarios where iHOP web services can be successfully used. This manuscript describes some of these scenarios, and the relevant changes applied to the family of iHOP web services.

2 Materials and Methods: iHOP Web Services Evolution

The need of systematically extracting information from literature in a number of biological projects has created a demand that has helped us to shape and extend the number of iHOP web services. The initial improvements focused on

A.T. Freitas and A. Navarro (Eds.): JBI 2010, LNBI 6620, pp. 102–107, 2012.
© Springer-Verlag Berlin Heidelberg 2012

service concurrency and scalability. We realized for instance that parallel queries as part of a massive study were putting an enormous pressure on the entire iHOP infrastructure. One of the solutions was setting up a gray lists system for recurring IPs in a short time range (for instance, a day) introducing increasing delay penalties. The other one was to limit the number of concurrent queries being attended, using a waiting queue, so internal databases (used by both web literature server and the web services) were not overloaded with queries.

The initial list of web services has grown in various ways and in the different web service categories. For instance, there were originally six REST[8] web services, based on basic iHOP functionality: related biological (gene or protein) symbols identification from free text; basic available information from a biological symbol identified by iHOP; fetch of abstract sentences used by iHOP system to model the definition of an identified biological symbol; relevant sentence look-up where a detected biological symbol is co-occurring with other ones; and fetch of iHOP annotated PubMed abstracts. Now there are over 20 REST web services, where some of them are refined versions of previous ones (getSymbolsFromSynonym, getSymbolsFromReference, guessSymbolId-FromReference, etc...), others are useful to third-party users (like redirect-FromSynonymToInteractions, for instance), and also new ones have been added (for instance, availableSymbolsFromTaxId, availableOrganisms or getLatestSymbolInformation). Many of these services have been created to attend specific needs of genome-wide analysis workflows.

There are also new iHOP SOAP services used by the FuncNet system (Clegg et al, submitted). FuncNet is a web-based tool for predicting when human proteins of unknown or poorly-understood function (a query set) are involved in the same processes or phenomena as proteins of a distinct and well-characterized function (a reference set). It is designed to help experimentalists narrow down large lists of proteins from high-throughput experiments to more tractable shortlists of candidates for individual assays. iHOP web services are part of the FuncNet statistical ensemble of algorithms hosted at various sites.

In the context of the EMBRACE consortium[1] the original iHOP SOAP web services described in WSDL as RPC/encoded services have been encoded in alternative approaches such document/literal, WS-I 1.0 Basic Profile compliant WSDL and are part of the EMBRACERegistry[10].

In the future the problems of web service interoperability at syntactic and semantic levels will be a key bottleneck for the usability of existing web services. In preparation for these new challenges iHOP web services have now an experimental port of iHOP XML Schema to BioXSD (an XML Schema with common simple and complex bioinformatic data type), using BioXSD wherever it is reasonable. iHOP XML Schema is used to define the representation of most iHOP web services responses in their different incarnations, so these annotations can be extended to the complete web services family. Additionally, we have also developed an experimental adaptation of document/literal WSDL definition of iHOP SOAP web services, so they have been semantically annotated using

EDAM ontology (EMBRACE Data and Methods ontology for bioinformatics tools and data) and SAWSDL[3] (technology which allows embedding semantic annotations on WSDL documents).

In the future, we plan to continue expanding the iHOP family of web services to facilitate the programmatic use of iHOP in large scale genome studies. Key technical issues related with web service operativity and scientific ones related with the estimation of probabilities of interaction in large interaction networks will require particular attention.

3   Use Cases

Two of the first use cases were the identification of spindle proteins literature mining (Rojas et al, submitted) and the construction of the back-end engine behind the iHOP widget accessible in the CARGO[5] web portal. In the following we describe in some detail two recent applications in the context of the ENFIN NoE[2] which combined the use of the iHOP web services with other bioinformatics and experimental approaches. These applications were successfully used for the identification of sub-network of potential interactors in two key biological processes, i.e. angiogenesis and late anaphase chromatin condensation.

3.1   Angiogenic Protein Sub-network

Angiogenesis is a major mechanism of vascularization during embryonic development, growth, formation of the corpus luteum and endometrium, regeneration and wound healing. Deregulated, abnormal angiogenesis is involved in pathological processes such as cancer, playing an essential role in tumor growth, invasion, and metastasis.

The project was carried out in collaboration with Juan A.G. Ranea and Francisca Sánchez from ProCel group - U. of Málaga, Jaak Vilo team from AS EGEEN, and Andrew Clegg and Christine Orengo from UCL. It involved the initial generation of a curated set of proteins, comprising all the reliable current angiogenic proteins found in GO and literature, and manually assessed by domain experts. 341 proteins were obtained, and this list of proteins was used as a seed to obtain other plausible candidates using different data mining and bioinformatics prediction methods [12].

iHOP web services were used to complement the information provided by those methods with an orthogonal approximation. iHOP web services provided information about sentences where two or more gene/protein symbols co-occur, providing specific information about the reliability of the gene/protein symbol detection, and the basic statistics to calibrate the reliability of the interactions about the number of co-occurrences for each protein/gene and the threshold of gene detection score found in those sentences. A total of 84 453 angiogenic target predicted associations were extracted from the literature, taking two days the whole process. 729 of them were considered highly reliable predicted pairs ($p\_val \leq 0.01$). These pairs were considered for experimental validation together with the ones predicted by other methods[11]. All the high quality predictions were integrated by the ProCel group in a single protein network which should represent a Human angiogenic protein sub-network (see Figure 1).

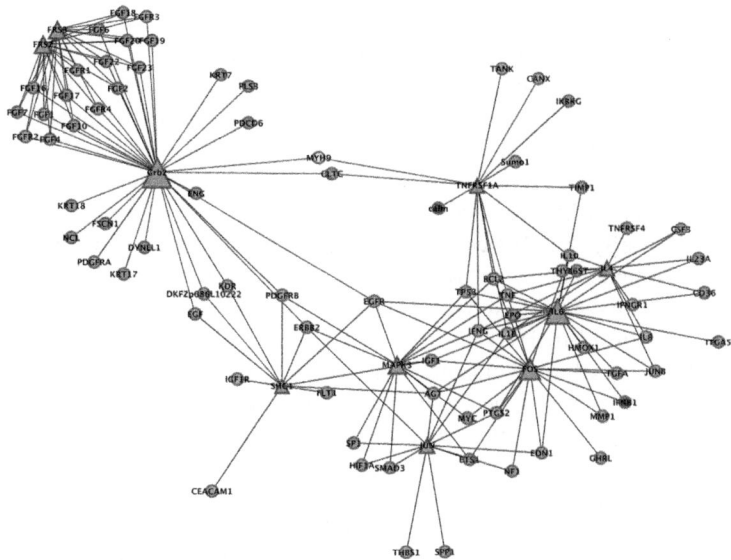

Fig. 1. Partial view of the integrated Angiogenesis sub-network. Triangle nodes represent known angiogenic proteins. Circles represent associated predicted targets. Figure taken from the public report delivered to ENFIN (http://www.enfin.org/)

3.2   Late Anaphase Chromatin Condensation

iHOP Web Services were also applied to the identification of proteins involved in mitotic chromosome condensation during late anaphase (LACC), as well as other potential regulator of chromatin architecture (RCA), in the context of a consortium of computational and experimental biologists interested in this problem. Those groups included Christine Orengo's group from UCL, Ana Rojas from CCBG-IMPPC, Thomas Skot Jensen group from CBS, Jaak Vilo's group from BIIT, Jean-Karim Heriche group from EMBL, Juan A.G. Ranea from U. of Málaga and our group at CNIO.

In the case of LACC the starting point was a small number of proteins known to interact with KIF22. KIF22 is a member of kinesin-like protein family. This family of proteins are microtubule-dependent molecular motors that transport organelles within cells and move chromosomes during cell division. Studies with the Xenopus homolog suggests its essential role in metaphase chromosome alignment and maintenance[6].

In this case the initial set of known proteins was too small, and the biological process itself insufficiently characterized at the experimental level. Therefore, it was impossible to find a significant number of co-occurrences in the literature. Therefore, we extended the iHOP search space to provided not only direct co-occurrences but also indirect ones, filtering those that provide a more clear evidence of physical/biological interaction. This process that naturally increased the capacity of the system to extract interactions at the expenses of decreasing their reliability.

In this case we took entire human gene set as a starting point, using as reference the HGNC[4] database subset of truly identified and accepted genes, which are better referenced in literature and their mapping is less problematic than sources like Ensembl. Each HGNC gene id was mapped to iHOP ids using the iHOP symbol identification service guessSymbolIdFromReference. When the mapping based on identifiers failed, we used each HGNC gene name and its synonyms with guessSymbolIdFromSynonym in order to increase recall. We hand-curated the obtained results to avoid possible misidentifications, because some databases, i.e. NCBI Entrez Gene, use plain number identifiers for human genes colliding the ones used in HGNC.

We extracted all the available sentences about each gene co-occurring with other human genes and pseudo-genes, containing a verb indicating true interactions (for instance, 'interact', 'bind' or 'complex'). In this case we used the internal iHOP score to select the most significant sentences. The level of reliability was established by a combination of quality of the sentences and likelihood of being a true gene/protein. In practice we divided the set of obtained sentences in direct and indirect co-occurrence sentences. In the first subset, genes co-occurring in the sentence are more likely to be truly HGNC identified genes, and were assigned a higher confidence value. In the second subset, the schema is A S1 U S2 B, where A and B are from HGNC set, U is a human gene not yet confirmed by HGNC (for instance, putative or pseudo-gene, so it is not included in the HGNC set), and S1 and S2 are sentences with physical verbs where U appears, A appears in S1 and B appears in S2. So the names in sentences with more evidence of interaction and better defined in the database were preferred to those with weaker sentences and less clearly described as genes/proteins.

With this strategy we obtained some evidences from literature about putative genes performing a role (related to physical participation) in LACC. These type of possible interactions, which describe the whole human genome physical interaction network from the literature point of view, are now assembled in general predictor that will gain from the synergy of orthogonal computational approaches.

Acknowledgments. We wish to thank Chris Sander and his group cBio@MSKCC (Computational Biological Center at Memorial Sloan-Kettering Cancer Center) for hosting the iHOP infrastructure, including the iHOP web server. Without their continuous support, iHOP could not be maintained publicly available. iHOP web services are supported by Spanish National Institute for Bioinformatics (www.inab.org), a platform of the Instituto de Salud Carlos III. Some of the works described here have been funded by ENFIN Network of Excellence (LSHG-CT-2005-518254) and EMBRACE Network of Excellence (LHSG-CT-2004-512092).

References

1. EMBRACE Grid Network of Excellence, http://www.embracegrid.info/
2. ENFIN Network of Excellence, http://www.enfin.org/
3. Semantic Annotations for WSDL Working Group,
   http://www.w3.org/2002/ws/sawsdl/
4. Bruford, E.A., Lush, M.J., Wright, M.W., Sneddon, T.P., Povey, S., Birney, E.:
   The HGNC database in 2008: a resource for the human genome. Nucl. Acids
   Res. 36(suppl_1), D445–D448 (2008)
5. Cases, I., Pisano, D.G., Andres, E., Carro, A., Fernández, J.M., Gómez-López, G.,
   Rodriguez, J.M., Vera, J.F., Valencia, A., Rojas, A.M.: CARGO: a web portal to
   integrate customized biological information. Nucleic Acids Research 35(Web Server
   issue), W16–W20 (2007); PMC1933121
6. Feine, O., Zur, A., Mahbubani, H., Brandeis, M.: Human kid is degraded by
   the APC/C(Cdh1) but not by the APC/C(Cdc20). Cell Cycle (Georgetown,
   Tex.) 6(20), 2516–2523 (2007); PMID: 17726374
7. Fernández, J.M., Hoffmann, R., Valencia, A.: iHOP web services. Nucleic Acids
   Research 35(Web Server issue), W21–W26 (2007); PMC1933131
8. Fielding, T.: Architectural Styles and the Design of Network-based Software
   Architectures. Ph.D. thesis, University of California, Irvine (2000),
   http://www.ics.uci.edu/~fielding/pubs/dissertation/top.htm
9. Hoffmann, R., Valencia, A.: A gene network for navigating the literature. Nat.
   Genet. 36(7), 664 (2004)
10. Pettifer, S., Ison, J., Kalas, M., Thorne, D., McDermott, P., Jonassen, I.,
    Liaquat, A., Fernandez, J.M., Rodriguez, J.M., Partners, I., Pisano, D.G.,
    Blanchet, C., Uludag, M., Rice, P., Bartaseviciute, E., Rapacki, K., Hekkelman,
    M., Sand, O., Stockinger, H., Clegg, A.B., Bongcam-Rudloff, E., Salzemann, J.,
    Breton, V., Attwood, T.K., Cameron, G., Vriend, G.: The EMBRACE web service
    collection. Nucl. Acids Res. 38(suppl_2), W683–W688 (2010)
11. Ranea, J., Morilla, I., Lees, J.G., Reid, A., Yeats, C., Clegg, A.B., Fernández, J.M.,
    Valencia, A., Sanchez-Jiménez, F., Orengo, C.: Angiogenic protein sub-network
    report for ENFIN. Personal Communication (October 2009)
12. Ranea, J., Morilla, I., Lees, J.G., Reid, A., Yeats, C., Clegg, A.B., Sanchez-Jiménez,
    F., Orengo, C.: "Dark Matter" assessment in protein network prediction and
    modelling. In: JBI 2010 Proceedings (October 2010)

Bioinformatic Software Developments in Spain

José M. Fernández[1] and Alfonso Valencia[2]

[1] GN2, Spanish National Bioinformatics Institute (INB)
[2] Spanish National Bioinformatics Institute (INB)
Structural Biology and Biocomputing Programme, CNIO, Spain
{jmfernandez,avalencia}@cnio.es

Abstract. The Spanish National Bioinformatics Institute (Instituto Nacional de Bioinformática in Spanish, or short INB) is an academic service institution founded in 2003 by the mayor research groups in Spain at that time. The INB serves in the coordination, integration and development of Spanish Bioinformatics Resources in projects in the areas of genomics, proteomics and translational medicine. Its mission is to consolidate Bioinformatics as a scientific discipline, providing technical support in Bioinformatics to laboratories, institutions and companies throughout the territory.

The JBI2010 conference featured two sessions, "INB Technicians internal session" and "Bioinformatic Software Developments in Spain and beyond", that introduced the state of the art of bioinformatic software developments at the INB and its role at the national and international level. This paper gives a summary of those sessions and presents an overview of the activities and contributions of the INB to the field of bioinformatics.

Keywords: Web services, Web tools, Pipeline analysis, Molecular Dynamics.

1 Introduction

Since the early days of bioinformatics many resources like biological databases or analysis tools have been accessible to the community via the Internet. In the beginning, methods like e-mail or FTP were used to run Blast searches or to access the raw content of databases. Later web browsers gained traction among the users due to their graphical capabilities and the possibility to easily link content. The first web applications were basically thin software layers built around bioinformatic algorithms and data retrieval systems focused on human interaction. Since then these systems have grown in number, complexity and quality and are now one of the cornerstones of biomedical research.

2 INB Bioinformatic Tools and Systems

At the time the term World Wide Web was coined web applications were either static pages or dynamic content generated on the server side, with limited navigation and interaction capabilities in both cases. Desktop applications were the

A.T. Freitas and A. Navarro (Eds.): JBI 2010, LNBI 6620, pp. 108–120, 2012.

only way to provide a rich, graphical and integrated experience. But in many cases it was difficult to run these systems locally due to their hardware requirements and the need to install large databases. As the different web technologies have become more powerful (web application servers, applets, Flash, AJAX, HTML5, WebGL) the distinction between desktop and web application is becoming less important. Rich Internet applications now provide a very similar experience to the user with the advantage that computationally expensive tasks can be performed on the server side making such systems more widely accessible. The INB plays a very active role in the development of end user web tools to assist in different areas of life sciences research. In this section we present some of these systems and discuss their specific contributions to Bioinformatics.

BlastXP[6] (BLAST Cross-Pharmacology) is a publicly accessible web tool for identifying proteins that are pharmacologically related through the number of shared ligands between them. BlastXP takes as input a protein sequence or database identifier and searches the WOMBAT database to extract the ligands of that protein. These ligands are then used to search for proteins with similar ligands. The results are ordered according to the 2D similarity of the ligands the proteins share.

The APPRIS[3] system deploys a range of computational methods to add additional value to the annotations of the human genome. In the curation procedure one of the CDS for each gene is selected as the principal functional unit combining protein structural information, functionally important residues and evidences of non-neutral evolution of exons amongst others. The server is used in the ENCODE project[15,32] to annotate 100% of the human genome (22,304 genes and 72,731 transcripts). APPRIS also includes a query-oriented data management system based on BioMart[17] to provide annotation-like searches of complex descriptive data and a simple visualization based on UCSC genome browser[22].

aGEM 2.0[20,2] is a powerful and intuitive platform that addresses the issue of the diversity and heterogeneity of anatomical gene expression. It integrates five mouse gene expression resources (EMAGE, Gene Expression Database, GENSAT database, Allen Brain Atlas database and BioGPS) and general biological information from OMIM, KEGG and MTB. MGI is used as the standard for genes and the EMAP/MA ontologies for the mouse embryo and adult animals (from Theiler Stage 1 to 28) have been considered as the standard for anatomical structures. The anatomical vocabularies from these databases have been aligned semi-automatically to the anatomical ontologies mentioned before. All this information is mapped into a unique integrative data model, combining information from genes and anatomical structures.

Babelomics[24,4] was created to respond to the growing necessity of integrating and analyzing different types of genomic data in an environment that allows an easy functional interpretation of the results. Babelomics is a complete suite of methods for the analysis of gene expression data that include normalization, pre-processing, differential gene expression, predictors, clustering and large-scale genotyping assays. All these genomic data analysis facilities are integrated and connected to multiple options for the functional interpretation of the experiments. Different methods of functional enrichment or gene set enrichment can be used to understand the functional basis of the experiment analyzed. Many sources of biological information were integrated that include functional information (GO, KEGG, Biocarta, Reactome, etc.), regulatory information (Jaspar, ORegAnno, miRNAs, etc.), results from text-mining methods or protein-protein interactions. Recently, methods for the de novo functional annotation of sequences has been included in the system.

Molecular dynamics is a very powerful tool that can be used to observe up to microsecond dynamics of proteins. However, the process requires a complex set-up which is often out of reach for scientists who are not experts in the field. To improve the access to this type of information the MoDEL[25] database with the trajectories of more than 1700 proteins was published. In this system a new workflow is being developed that will integrate existing tools into a modern platform to run molecular dynamics simulations. This will allow users without any previous knowledge to run MD simulations and analyze the trajectories without the need of an expert.

Projects like Chronic Lymphocytic Leukemia - International Cancer Genome Consortium (CLL-ICGC[7]) or ESP-SOL (tomato sequencing project) have generated huge volumes of experimental information. Analysis like genome assembly, copy number variation calculation, mutation, SNPs and InDels detection require more and more the support of supercomputing facilities. Simulations of molecular dynamics and protein docking are other scenarios where large computational resources that go beyond the capacity of individual research centers are necessary. The Barcelona Supercomputing Center is part of the INB and provides access to the MareNostrum system that makes is possible to address these challenges. Some examples of these intensive algorithms are high-throughput docking, py-DockSAXS, SIPPER, FTDdock-CELL, pyDockWEB and Ensemble docking. In these cases, the size of the inputs (biological sequences) is very small, compared to the huge computational needs and the size of the intermediate results. In comparison, the reads from next-generation sequencing experiments present in themselves already large volumes of data that have to be processed.

Genome-wide Association Studies (GWAS) based on Single Nucleotide Polymorphism (SNP) arrays are another example of studies that produce large data volumes. Currently, the complexity of the methods and software packages available and the proliferation of custom formats makes it difficult for scientists to process this kind of data. To address these problems GWASpi[10] was developed as a user-friendly, multi-platform, desktop software for the management and analysis of GWAS data. It uses Array-oriented Scientific Data Format databases (NetCDF) as part of an effort to solve the increasing problems of data storage and manipulation. Because of the achieved leap of performance GWASpi now enables "start to end" GWAS management, from raw data to end results and charts, on commonly available desktop hardware.

3   Web Services

Web applications are designed for human interaction, but as more and more integration of different systems is desired, web services have become popular in bioinformatics to access tools programmatically. The coordinated usage of public web services has allowed ever more complex analysis over growing bioinformatic data sets. Experience has shown that web services are a key technology to apply heterogeneous bioinformatic methods in a massively distributed analysis.

Web services are conceptually similar to remote procedure calls: the service receives a request with the query from a remote entity (an user, a program); the query is decoded from the transport format to native formats; the algorithm implemented in the service performs its computations and generates the results; these results are encoded from native formats to the transport format; the result is then returned to the remote entity.

There have been several distributed service paradigms in use in bioinformatics: REST, SOAP, Soaplab[29], BioMOBY[33], DAS[19], CORBA, etc. They share common principles but differ on features like message representation or service description. Some of these protocols, like DAS, Soaplab or BioMOBY, were originally designed for the bioinformatics community, while others were developed for the integration of business processes or in the academia.

The proliferation and widespread adoption of web services made it almost impossible to know which web services were already available, and if these services were interoperable. This led to the creation of the first resource registries in bioinformatics, which were at the beginning lists of hyperlinks to the available tools, databases and web services with embedded descriptions. These lists evolved to specialized web services built around a databases that stores all this information, like BioMOBY Central (repository of BioMOBY web services, input, output and service descriptions), DASRegistry[27] and BioCatalogue[14] (a generic web services catalog).

Later, generic web service workflow engines like Taverna[18] or Triana[31] were developed to allow a neutral, graphical representation and design of workflows. These workflows are basically recipes that describe which web services are to be called and in which order, the translations to apply to the inputs before calling a web service, and which web service output is the final result. To share workflows and increase reuse workflow repositories like myExperiment[16] were created.

3.1   INB Web Services

Since its inception the Spanish National Bioinformatics Institute has been working with web service technologies to effectively integrate systems and tools developed in different INB nodes that are distributed over the Spanish territory. The INB has a portfolio of more than 700 interoperable web services (Figure 1) like sequence analysis, biochemistry, structural studies, text mining, etc. that allow to build complex analysis workflows. The web services paradigm used in most INB web services has been BioMOBY, and the INB BioMOBY Central catalog

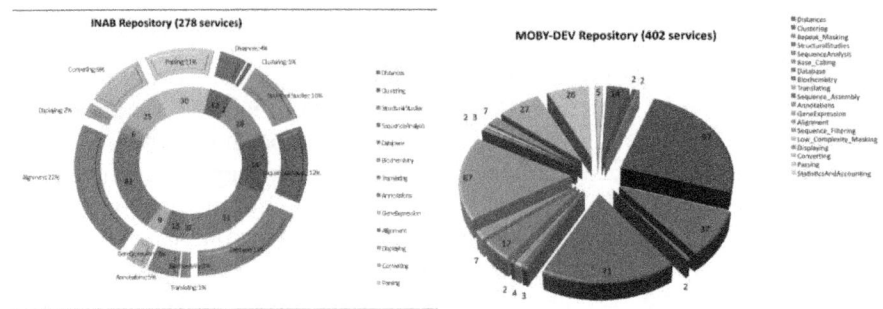

Fig. 1. Distribution of INB MOBY web services by service type

stores information about all the registered BioMOBY web services, their inputs, their outputs, the service descriptions and the kind of identifiers understood by each of the services.

All the BioMOBY services from a BioMOBY Central catalog share a common object ontology, used to describe the inputs and the outputs. This makes different web services interoperable with almost no supplementary integration work. All services are periodically tested using both naïve and predictable queries, in order to control their availability and stability. Also, a coordinated usage log of these services is maintained (see Figure 2).

The INB was involved in the EMBRACE Network of Excellence due its expertise in web technologies and participated in several BioHackathon[21] events. One of the objectives of EMBRACE was the exploration and optimization of information technologies usage in molecular biology and life sciences. The work on EMBRACE was focused on the integration of biological databases and bioinformatic software tools at a distributed level, which is one of the main problems bioinformatics, molecular biology, and in general life sciences suffer.

The EMBRACE Registry[26] was created inside EMBRACE as a testbed for some of the technologies related to web service catalogs: support the registration of web services using different protocols; web service look-up; scheduled service testing and validation using custom programs written and registered by each author; notification of service state changes to service developers, based on the results from service tests etc. Currently only some of these features are being reused in the BioCatalogue.

3.2   INB Systems for Working with Web Services

Scientific data analysis usually involves the combination of information provided by different data sources and tools. The proliferation of bioinformatic resources has created a great diversity of resource types that use different data representations and protocols. This heterogeneity of resources increases the complexity of data integration and is a burden for tool interoperability.

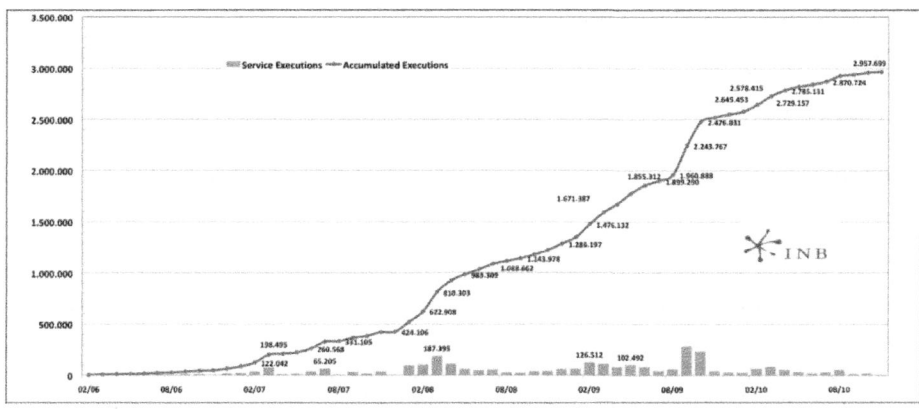

Fig. 2. Usage of INB MOBY web services, from January to November 10th, 2010

The Modular-API[13] (mAPI) is a programmatic framework that offers a uniform representation of web services implemented in different protocols to promote the integrated use of a diversity of bioinformatics resources.

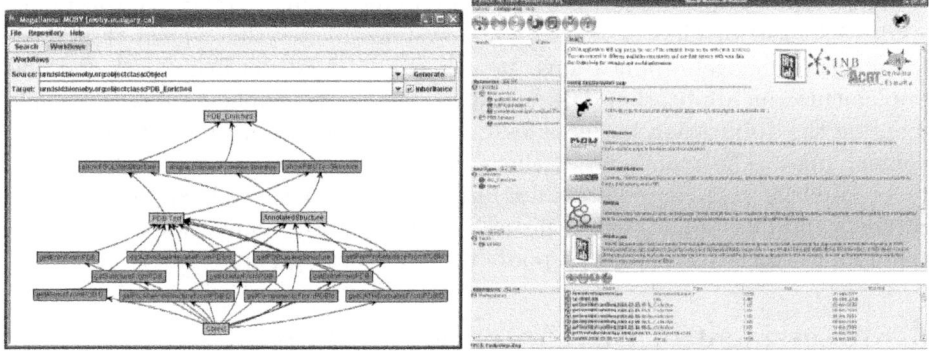

Fig. 3. Magallanes and jOrca in action, both using mAPI

Magallanes[28] is a powerful search engine specialized in web service resources discovery that explores all the available meta-data descriptions of the resources through syntactic and regular-expressions search methods. This facilitates the syntactic translation between inputs and outputs of web services. jORCA[23] facilitates the invocation of web services, using the uniform programmatic representation offered by mAPI and the discovery and composition of workflows supplied by Magallanes, providing a graphical representation for common bioinformatic data formats. Figure 3 contains a couple snapshots of these tools.

BioNemus[5] is a Java based platform for the development of semantic web services based on different ontology models. BioNemus provides a strong integration with the BioMoby platform as well as with EMBRACE ontology (EDAM[8]) to describe web service functionality. On the same line, the MobyMiner[12] applet is a vivid graphical, biological information harvesting platform based on BioMoby web services. The platform is implemented as a three-tier architecture with a Java based visualization tool working with a biological database over a set of web services.

The Interactive Web Workflow Enactor & Manager[11] (IWWE&M) has been created to enable the direct execution of workflows (both from the INB and my-Experiment) without the need of using additional systems. It addresses the issues of workflow dependencies and management and allows the execution of pre-stored workflows and the recollection and compilation of intermediate and final workflow enactment results. Programmers can directly integrate any in-course workflow enactment into their custom end user interfaces.

4   Clinical Data Management

Clinical histories from chronic patients are a very valuable, but at the same time complicated, source of knowledge that are obtained through periodical surveys from the patients and their relatives. The Form Builder (Figure 4) is a tool to compose and use custom forms or questionnaires in survey related projects, like epidemiological or clinical studies. The system follows the typical three layer architecture for web applications. It uses a relational database back-end that implements the application data model, a Java-based application server implementing the application business logic and a front-end implementing views and control rules. The main aim of the application is to be user friendly enough to be used by non computer-skilled users in a short time. It supports several projects and questionnaires per project, as well as user management. Currently, the application is installed to support three research groups.

5   Assessment of Bioinformatic Methods

The periodic assessment of existing systems and methods is an important task as it informs developers and users about their relative performance and how the field progresses. The assessed methods can be either experimental or computational in nature, and the assessors usually curate standard datasets for these assessments, focused on objective comparison of the results.

One of these assessment systems is the Electro Microscopy Benchmark (3DEMBenchmark[1]) that provides the infrastructure to dynamically evaluate several algorithms using a variety of datasets and a wide range of analysis tools. Developers can upload their results in a standardized format using data supplied by the web page (they can also download the input data for their private tests). The second kind of users are structural biologists who can analyze the available algorithms and decide which one better suits their needs.

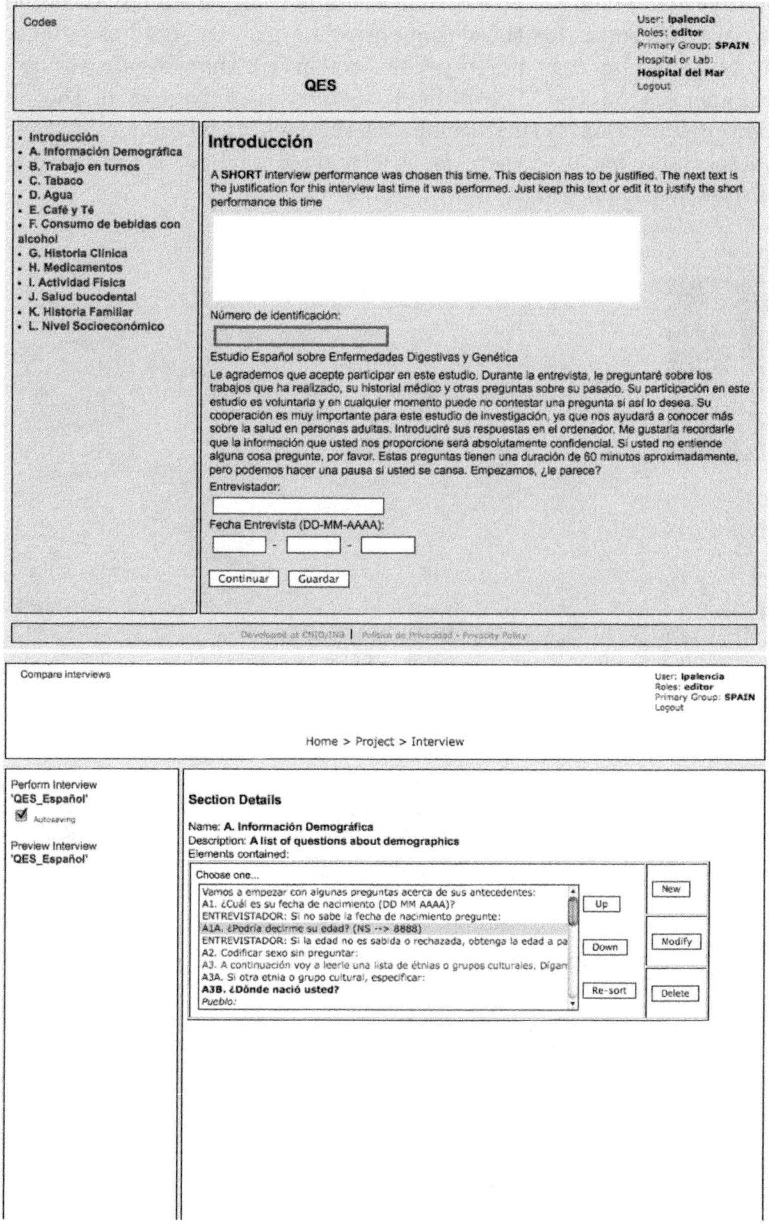

Fig. 4. Form Manager at the beginning of a clinical survey

Genome sequencing projects and NGS technology have found more novel sequences than can be annotated by manual efforts and the only viable way to obtain a functional characterization for them is to use automated methods. The prediction of GO terms for those sequences is a growing field as evinced by the numbers of Gene Ontology based prediction servers that are already working or in development. However, it is difficult to measure advances in the automatic prediction of functional terms because of the lack of reliable gold-standard annotation sets and because there is no standard evaluation framework for the function prediction protocols.

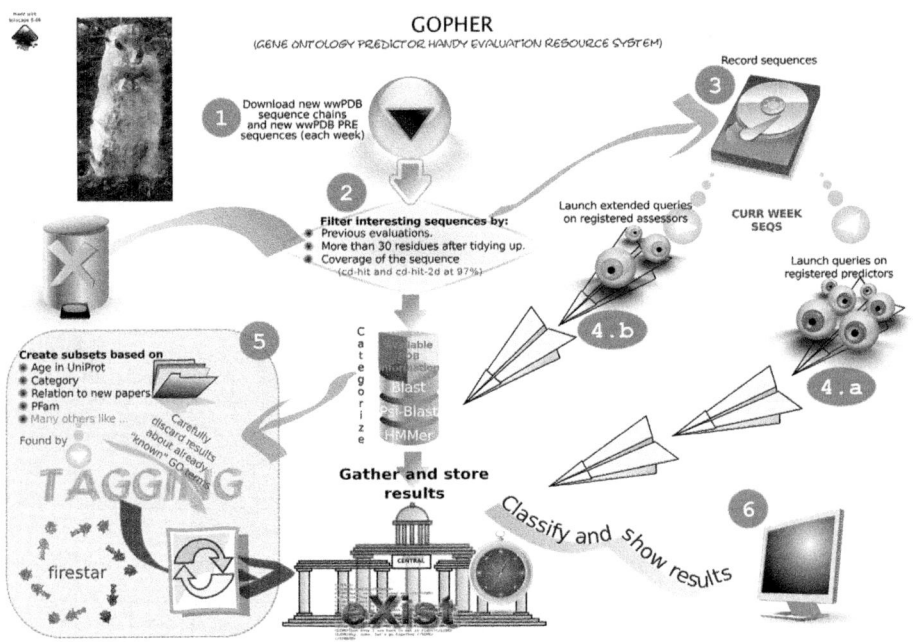

Fig. 5. GOPHER assessment workflow

The prediction of GO terms was included in the CASP experiment for the first time as part of CASP6[30]. However, the assessments carried out in the framework of the CASP6 and CASP7 experiments were hampered by the slow release of functional information, something that is not ideal for a rapidly developing field where predictors need to make use of the results and the evaluation in order to refine their methods. The need to organize a more effective blind function prediction assessment was obvious after CASP7.

GOPHER[9] was developed to assess the prediction of function on a continuous basis. The server obtains new structures from PDB (Figure 5) and prediction servers (i.e. participants) are invited to predict GO terms for those sequences for which function is not fully known. GOPHER then assesses the prediction of GO Molecular Function, Cellular Component and Biological Process terms. The assessment is automatic, but takes place some time after the release of the sequences because insights into the function of newly characterized protein structures are usually not made available immediately. In addition, the evaluation is revisited periodically because functional annotations can change over time.

6 Software Licensing Issues in Life Sciences

FreeBIT, the Iberoamerican Network for Free and Open Source Software in the Life and Health Sciences, funded by CYTED, was introduced in a talk and the main aspects of free software development were presented. The role of the FLOSS (Free/Libre/Open Source) development and its use in academic environments and its advantages for resource savings and knowledge sharing compared to open, closed and mixed source business models were reviewed. Different issues associated with the main Open Source licenses and their relative advantages for efficient resource usage and knowledge sharing in the academic world like public domain, BSD, GPL and EU-GPL and the use of Creative Commons licenses for other works were discussed.

7 Discussion and Conclusions

The Spanish National Bioinformatics Institute has been one of the first European networks to coordinate national developments and research initiatives in Bioinformatics. Since its inception it has been active in the development of tools and analysis systems, it was one of the drivers in the introduction of web services for system integration and it has been active in training initiatives and providing support to many sequencing projects. Here we have discussed some of the areas of activity of the INB and given a summary of the sessions at JBI2010 dedicated to the INB. For more information visit our web site `http://www.inab.org/`.

As we have seen, web tools, web services, and in general, software developments, are becoming the de-facto way knowledge is stored and made accessible to researchers and the wider public. But many of those systems are not maintained (especially smaller and mid-sized projects) because they lack stable funding. This leads inevitably to the loss of important and useful resources and diminished return of publicly funded research. To remedy this situation in the future the ELIXIR infrastructure is being developed of which the INB will be a part. The mission of ELIXIR will be to construct and operate a sustainable infrastructure for biological information in Europe to support life science research and its translation to medicine and the environment, the bio-industries and society. The purpose of ELIXIR is to build a plan for a sustainable infrastructure for biological information in Europe. This plan focuses on generating stable funding for Europe's most important publicly accessible databases of molecular biological information, and the development of a compute infrastructure that can cope with the biological data deluge. For more information visit `http://www.elixir-europe.org/`.

Acknowledgments. We wish to thank all these sessions participants and INB staff, who have provided descriptions of their servers and tools. Spanish National Institute for Bioinformatics (INB) is a platform of the Instituto de Salud Carlos III, and all the software and servers described above, but EMBRACERegistry, BioCatalogue, DASRegistry, Taverna and Triana, have been developed by its staff. Some of the works described above have also been funded by ENFIN Network of Excellence (LSHG-CT-2005-518254) and EMBRACE Network of Excellence (LHSG-CT-2004-512092).

References

1. 3DEMBenchmark, http://ipc.cnb.csic.es/3dembenchmark/
2. aGEM, http://agem.cnb.csic.es/
3. APPRIS, http://appris.bioinfo.cnio.es/
4. Babelomics, http://www.babelomics.org/
5. BioNemus, http://inb.bsc.es/documents/java/nemus/index.html
6. BlastXP, http://cgl.imim.es:8080/blastxp/
7. CLL-ICGC page, http://www.icgc.org/icgc/cgp/64/530/826
8. EMBRACE Ontology for Data and Methods (EDAM), http://edamontology.sourceforge.net/
9. GOPHER, http://gopher.bioinfo.cnio.es/
10. GWASpi, http://www.gwaspi.org
11. IWWE&M, http://ubio.bioinfo.cnio.es/biotools/IWWEM/
12. MobyMiner, http://inb.bsc.es/applications/java/mobyminer/moby_miner.html
13. ModularAPI, http://chirimoyo.ac.uma.es/mapi/
14. Bhagat, J., Tanoh, F., Nzuobontane, E., Laurent, T., Orlowski, J., Roos, M., Wolstencroft, K., Aleksejevs, S., Stevens, R., Pettifer, S., Lopez, R., Goble, C.A.: BioCatalogue: a universal catalogue of web services for the life sciences. Nucleic Acids Research 38(Web Server), W689–W694 (2010), http://www.nar.oxfordjournals.org/cgi/doi/10.1093/nar/gkq394
15. The ENCODE Project Consortium: Identification and analysis of functional elements in 1% of the human genome by the ENCODE pilot project. Nature 447(7146), 799–816 (2007), http://dx.doi.org/10.1038/nature05874
16. Goble, C.A., Bhagat, J., Aleksejevs, S., Cruickshank, D., Michaelides, D., Newman, D., Borkum, M., Bechhofer, S., Roos, M., Li, P., Roure, D.D.: Myexperiment: a repository and social network for the sharing of bioinformatics workflows. Nucleic Acids Research 38(Web Server), W677–W682 (2010), http://nar.oxfordjournals.org/content/38/suppl_2/W677.full?sid=b5dfd4bc-f807-49eb-acc1-09e39c218f81
17. Haider, S., Ballester, B., Smedley, D., Zhang, J., Rice, P., Kasprzyk, A.: BioMart central portal—unified access to biological data 37(Web Server issue), W23–W27 (2009); PMID: 19420058 PMCID: 2703988
18. Hull, D., Wolstencroft, K., Stevens, R., Goble, C., Pocock, M.R., Li, P., Oinn, T.: Taverna: a tool for building and running workflows of services. Nucleic Acids Research 34(Web Server), W729–W732 (2006), http://nar.oxfordjournals.org/content/34/suppl_2/W729.full

19. Jenkinson, A.M., Albrecht, M., Birney, E., Blankenburg, H., Down, T., Finn, R.D., Hermjakob, H., Hubbard, T.J., Jimenez, R.C., Jones, P., Kähäri, A., Kulesha, E., Macías, J.R., Reeves, G.A., Prlić, A.: Integrating biological data – the distributed annotation system. BMC Bioinformatics 9(suppl. 8), S3 (2008), http://www.pubmedcentral.nih.gov/articlerender.fcgi?artid=2500094, PMC2500094

20. Jiménez-Lozano, N., Segura, J., Macías, J.R., Vega, J., Carazo, J.M.: aGEM: an integrative system for analyzing spatial-temporal gene-expression information. Bioinformatics 25(19), 2566–2572 (2009), http://bioinformatics.oxfordjournals.org/content/25/19/2566.abstract

21. Katayama, T., Arakawa, K., Nakao, M., Ono, K., Aoki-Kinoshita, K.F., Yamamoto, Y., Yamaguchi, A., Kawashima, S., Chun, H.-W., Aerts, J., Aranda, B., Barboza, L.H., Bonnal, R.J., Bruskiewich, R., Bryne, J.C., Fernández, J.M., Funahashi, A., Gordon, P.M., Goto, N., Groscurth, A., Gutteridge, A., Holland, R., Kano, Y., Kawas, E.A., Kerhornou, A., Kibukawa, E., Kinjo, A.R., Kuhn, M., Lapp, H., Lehvaslaiho, H., Nakamura, H., Nakamura, Y., Nishizawa, T., Nobata, C., Noguchi, T., Oinn, T.M., Okamoto, S., Owen, S., Pafilis, E., Pocock, M., Prins, P., Ranzinger, R., Reisinger, F., Salwinski, L., Schreiber, M., Senger, M., Shigemoto, Y., Standley, D.M., Sugawara, H., Tashiro, T., Trelles, O., Vos, R.A., Wilkinson, M.D., York, W., Zmasek, C.M., Asai, K., Takagi, T.: The DBCLS BioHackathon: standardization and interoperability for bioinformatics web services and workflows. Journal of Biomedical Semantics 1(1), 8 (2010), http://www.jbiomedsem.com/content/1/1/8

22. Kent, W.J., Sugnet, C.W., Furey, T.S., Roskin, K.M., Pringle, T.H., Zahler, A.M., Haussler, D.: The human genome browser at UCSC. Genome Research 12(6), 996–1006 (2002), http://genome.cshlp.org/content/12/6/996.abstract

23. Martín-Requena, V., Ríos, J., García, M., Ramírez, S., Trelles, O.: jORCA: easily integrating bioinformatics web services. Bioinformatics 26(4), 553 (2010), http://bioinformatics.oxfordjournals.org/content/26/4/553.abstract

24. Medina, I., Carbonell, J., Pulido, L., Madeira, S.C., Goetz, S., Conesa, A., Tarraga, J., Pascual-Montano, A., Nogales-Cadenas, R., Santoyo, J., Garcia, F., Marba, M., Montaner, D., Dopazo, J.: Babelomics: an integrative platform for the analysis of transcriptomics, proteomics and genomic data with advanced functional profiling. Nucleic Acids Research 38(Web Server), W210–W213 (2010), http://nar.oxfordjournals.org/content/38/suppl_2/W210.long

25. Meyer, T., D'Abramo, M., Hospital, A., Rueda, M., Ferrer-Costa, C., Pérez, A., Carrillo, O., Camps, J., Fenollosa, C., Repchevsky, D., Gelpí, J.L., Orozco, M.: MoDEL (Molecular dynamics extended library): A database of atomistic molecular dynamics trajectories. Structure 18(11), 1399–1409 (2010), http://www.sciencedirect.com/science/article/B6VSR-51F2C49-5/2/9a769c5e76c6bc241336bf53de30d586

26. Pettifer, S., Ison, J., Kalas, M., Thorne, D., McDermott, P., Jonassen, I., Liaquat, A., Fernandez, J.M., Rodriguez, J.M., Partners, I., Pisano, D.G., Blanchet, C., Uludag, M., Rice, P., Bartaseviciute, E., Rapacki, K., Hekkelman, M., Sand, O., Stockinger, H., Clegg, A.B., Bongcam-Rudloff, E., Salzemann, J., Breton, V., Attwood, T.K., Cameron, G., Vriend, G.: The EMBRACE web service collection. Nucl. Acids Res. 38(suppl_2), W683–W688 (2010)

27. Prlic, A., Down, T., Kulesha, E., Finn, R., Kahari, A., Hubbard, T.: Integrating sequence and structural biology with DAS. BMC Bioinformatics 8(1), 333 (2007), http://www.biomedcentral.com/1471-2105/8/333

28. Rios, J., Karlsson, J., Trelles, O.: Magallanes: a web services discovery and automatic workflow composition tool. BMC Bioinformatics 10(1), 334 (2009), http://www.biomedcentral.com/1471-2105/10/334

29. Senger, M., Rice, P., Oinn, T.: Soaplab - a unified sesame door to analysis tools. Proc UK e-Science programme All Hands Conference 6, 2–4 (2003), http://citeseerx.ist.psu.edu/viewdoc/summary?doi=10.1.1.66.4856

30. Soro, S., Tramontano, A.: The prediction of protein function at CASP6. Proteins: Structure, Function, and Bioinformatics 61(S7), 201–213 (2005), http://dx.doi.org/10.1002/prot.20738

31. Taylor, I., Shields, M., Wang, I., Harrison, A.: The triana workflow environment: Architecture and applications. In: Workflows for e-Science, pp. 320–339 (2007), http://dx.doi.org/10.1007/978-1-84628-757-2_20

32. Tress, M.L., Martelli, P.L., Frankish, A., Reeves, G.A., Wesselink, J.J., Yeats, C., Ísólfur Ólason, P., Albrecht, M., Hegyi, H., Giorgetti, A., Raimondo, D., Lagarde, J., Laskowski, R.A., López, G., Sadowski, M.I., Watson, J.D., Fariselli, P., Rossi, I., Nagy, A., Kai, W., Størling, Z., Orsini, M., Assenov, Y., Blankenburg, H., Huthmacher, C., Ramírez, F., Schlicker, A., Denoeud, F., Jones, P., Kerrien, S., Orchard, S., Antonarakis, S.E., Reymond, A., Birney, E., Brunak, S., Casadio, R., Guigo, R., Harrow, J., Hermjakob, H., Jones, D.T., Lengauer, T., Orengo, C.A., Patthy, L., Thornton, J.M., Tramontano, A., Valencia, A.: The implications of alternative splicing in the ENCODE protein complement. Proceedings of the National Academy of Sciences 104(13), 5495–5500 (2007), http://www.pnas.org/content/104/13/5495.abstract

33. Wilkinson, M.D., Senger, M., Kawas, E., Bruskiewich, R., Gouzy, J., Noirot, C., Bardou, P., Ng, A., Haase, D., de Andres Saiz, E., Wang, D., Gibbons, F., Gordon, P.M.K., Sensen, C.W., Carrasco, J.M.R., Fernández, J.M., Shen, L., Links, M., Ng, M., Opushneva, N., Neerincx, P.B.T., Leunissen, J.A.M., Ernst, R., Twigger, S., Usadel, B., Good, B., Wong, Y., Stein, L., Crosby, W., Karlsson, J., Royo, R., Párraga, I., Ramírez, S., Gelpi, J.L., Trelles, O., Pisano, D.G., Jimenez, N., Kerhornou, A., Rosset, R., Zamacola, L., Tarraga, J., Huerta-Cepas, J., Carazo, J.M., Dopazo, J., Guigo, R., Navarro, A., Orozco, M., Valencia, A., Claros, M.G., Pérez, A.J., Aldana, J., Rojano, M.M., Cruz, R.F., Navas, I., Schiltz, G., Farmer, A., Gessler, D., Schoof, H., Groscurth, A.: Interoperability with moby 1.0–it's better than sharing your toothbrush! Briefings in Bioinformatics 9(3), 220–231 (2008), http://bib.oxfordjournals.org/cgi/content/abstract/9/3/220, PMID: 18238804

Author Index